ADVANCE PRAISE FOR LED BY LOVE OF COUNTRY

"If you want to write your father's story about his time and front-line service in World War II, you should do it the way David Miles has done it here in this book. Nowhere is the bond between father and son more visible than in this book. David Miles has created a masterpiece here, not only to memorialize his father, but also to remember the "greatest generation" of his father's comrades, especially those who gave their lives on the battlefield for our freedom. Also, those who gave their best to free Europe from the yoke of Nazism and were fortunate enough to return home. Thank you for this great book!"

— Albert Trostorf, Mayor of Merode, Germany, and author of
Lest We Forget, a seven-volume series on the Battle of the Hurtgen Forest

"Author David Miles lovingly presents readers with a very personal account of his father's tour of duty in the European Theater Operations based on his father's meticulous notes, memoirs, photos, and memorabilia collected during arduous service. Pfc. Ralph Miles makes you feel like you're standing beside him as he navigates the endless journey of a replacement, the frigid terror of the Hurtgen Forest, the determined defense of Luxembourg and the exhausting push through Southern Germany. Along the way the reader will come to feel attached to the heroic soldiers, like Miles, who suffered through the horrors of global warfare."

— Craig Chapman, historian and author of
Battle Hardened: An Infantry Officer's Harrowing Journey from D-Day to VE Day

"My grandfather served in the 4th Infantry Division during the Second World War. I never had the opportunity to discuss his experiences with him. This is the book I wish I had about his travels. Part memoir, part travelogue, part artifact book, this examination of a common GI's exploits across Europe is deeply personal. David Miles' introspective journey chronicling his father's wartime odyssey is a stirring example of how one can revive and contextualize family history."

— Jared Frederick, PhD., professor of History at the University of Pennsylvania-Altoona
and author of *Dispatches of D-Day*

Led by Love of Country

Led by Love of Country

A Journey through the WWII Journal of a 12th Regiment Radio Man

Ralph J. Miles with David W. Miles

Published by Deeds Publishing in Athens, GA
www.deedspublishing.com

Printed in The United States of America

Cover and interior design by Deeds Publishing

ISBN 978-1-961505-18-6

Books are available in quantity for promotional or premium use. For information, email info@deedspublishing.com.

First Edition, 2024

10 9 8 7 6 5 4 3 2 1

*"Behind every soldier, there is an even stronger woman who stands
behind him, supports him, and loves him with all her heart."*

Dedicated to my mother, Mary Eleanor Hobbs Miles, and to millions of others like her who counted the hours, the days, the months and the years until their soldiers came home — and especially to those thousands of women whose soldiers never came home.

Above: Shoulder patch of the Fourth Infantry Division
Motto: Steadfast and loyal

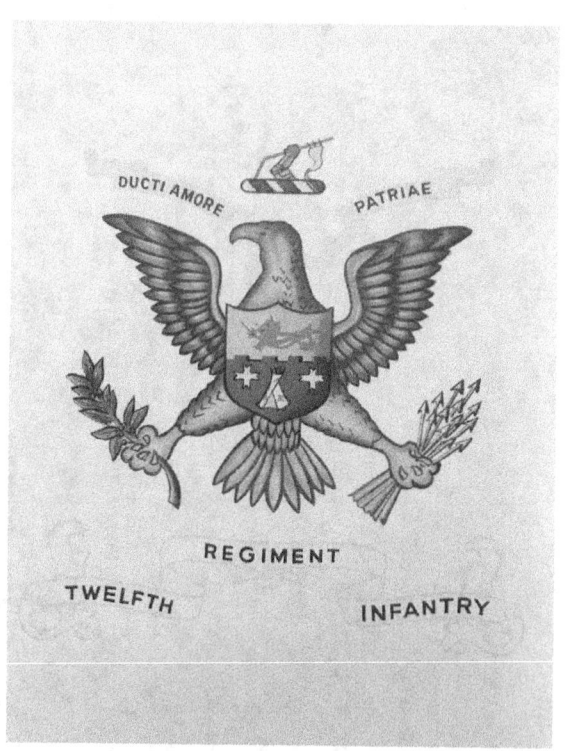

Above: Insignia of the 12th Infantry Regiment
Motto: Having Been Led by Love of Country

Contents

Private Ralph Jacob Miles, Senior, September 1943

Acknowledgements

A book such as this one cannot possibly happen without many people giving freely of their time, talents, wisdom and support. I cannot adequately express my indebtedness to each of the individuals listed on this page for their generous and indispensable contributions to this project. They patiently answered countless questions, allowed the use of personal material about their own ancestors, offered critical advice and guidance regarding the organization and presentation of this narrative, and have buoyed me with their constant encouragement and support whenever my own energy level began to flag. To each of them, I say a fervent, "*Thank you!*"

Michael Belis, veteran and WW II historian, with a particular interest in the 4th Infantry Division. FB page: Michael Belis.

Craig Chapman, veteran and author of many widely read works of historical non-fiction. Craig is the son of William P. "Bill" Chapman, who served as Captain of Company E, and as an S-3 at 2nd Battalion Headquarters, 12th Regiment, 4th Infantry Division, during World War II.

Leonard Cizewski, World War II historian with special expertise in the 522nd Field Artillery Battalion, a highly decorated Nisei unit attached to Task Force Rodwell and the 12th Infantry in the spring of 1945. Leonard has done extensive research in cooperation with other historians on the liberation of Dachau slave labor inmates by the 522nd FAB during the War.

Dr. Jared Frederick, Instructor of History at Penn State University-Altoona and author of numerous books concerning World War II, the Civil War and other American history topics.

Ansley Miles, my ten-year-old granddaughter, who suggested the title for her great-grandfather's book; thank you for your love and creative contribution, Rosebud!

Nancy Miles, my beautiful, supportive and talented wife, who contributed her considerable talents and many hours of her time as my personal copy editor, content advisor and literary critic. More than this, her love, her patience, her beauty and her character have truly made her a gift from God in my life.

Olaf Nitsch, an extraordinarily knowledgeable amateur World War II historian — FB page: Olaf Nitsch

Beth Reiman, author of *Unpacking Yesterday: Brotherhood's Legacy — Teddy Roosevelt, Jr., James S. Rodwell and the 4th Infantry Division in WWII*, for her invaluable assistance in identifying some of the photos Ralph took during his trip from Bamberg, Germany, to Le Havre, France in July of 1945.

Albert Trostorf, mayor of Merode, Germany and author of a seven-volume history of the 9th Infantry Division in the Battle of the Hurtgen Forest entitled, *Lest We Forget*. Albert is founder of *Arbeitz Gruppe Grenzland 1944-1945* (Borderlands Work Group 1944-1945), which hosts an annual fall conference on the Hurtgen battles and promotes historical understanding and reconciliation between the descendants of Allied and German combatants.

Leslie Weiner, daughter of Sergeant Henry C. Strecker of Rifle Company C, First Battalion, 12th Regiment, Fourth Infantry Division, for graciously allowing me to use information and mementos that she inherited from her father to supplement Ralph's recollections in his memoirs.

Co-Author's Note

Ninety-nine percent of this book consists of my father's own words, and a similar proportion of the cartoons, maps and pictures contained herein, he himself either clipped from his issues of *Stars and Stripes* and *Yank* Magazine, or he snapped with a buddy's camera. Dad was an excellent writer — concise, entertaining, informative. Since I can't possibly improve on the original, the reader will be grateful to know that I didn't begin to try.

However, as Dad wrote more than once, he knew little about the "big picture" or grand strategy of the war. He hardly knew where he was or where he had been, let alone where he was going or the reasons behind any of it. That is why he kept a diary, and when he had the chance, "liberated" kids' geography books along the way and devoured every military news sheet he could find, in an attempt to piece together what he could of his whereabouts and movements. Dad's limited view of things inevitably resulted in some gaps in the military picture. To remedy this, I have inserted at strategic points in the book supplementary information, maps and photos intended to provide the reader with some clarifying context and background for Dad's very personal stories.

I am sure that the reader will easily discern my photos, maps and captions from those in Dad's original memoirs without any explanation here. I did, though, set apart some of my contextual commentary with a line of asterisks (*****).

Ralph spent the years between 1945 and 1950 entering his war memories and memorabilia into a single volume. That album crumbled and frayed badly with much handling over the ensuing twenty years. In 1972, he painstakingly transferred his narrative and other items into two, better quality volumes.

I have vacillated many times over whether to donate his memoirs to the National Archives to become a part of the historical record, and thus accessible to historians, researchers and casual readers, or keep them as a priceless family heirloom. I decided to do both: publish this book for all to share in Ralph's personal history of a momentous national event, while keeping the originals in the hands of a family that has treasured them over the past eighty years — and I pray, for many more generations to come.

— *David Wayne Miles*

"To those who fought for it, freedom has a flavor the protected will never know."
— **P. McRee Thornton,** *The Star-Spangled Son*

Preface

On December 8, 1941, hundreds of thousands of the flower of American youth flooded armed forces recruiting offices across the United States to volunteer to become soldiers, sailors or marines. The malevolent forces of tyranny and subjugation, driven by twisted ideologies of ultra-nationalism, hatred and conquest, had attacked the nation from both the east and the west. That attack shook from its slumbers a here-to-fore disinterested and comfortably isolated country, and transformed it into an implacable, and enraged force that would not stop until it had ground its arrogant foes into dust.

Over the next four years, sixteen million American men and women answered their country's call and marched or sailed into the bloody maw of a catastrophic world war. Over one million would become casualties, including nearly 300,000 killed in action, consecrating with their blood the soil of countries most of them had hardly known existed.

One of those sixteen million was my father, Ralph Jacob Miles, Senior, one of seven children born to a hard-drinking, hard-working South Georgia sharecropper, Daniel Dozier Miles, and his long-suffering wife, Mary Jane "Janie" Rushing Miles. Like most of his generation, Ralph had little in the way of money, but much in the way of mental and physical toughness forged in the twin crucibles of grinding poverty and world war. He and his fellow GIs faced down both of these daunting foes with stoic and unflinching resolve. And because they did so, we today are a free, prosperous and secure people.

We owe these heroes — a designation they vehemently eschewed — a debt that defies repayment. To quote the famous line from Steven Spielberg's 1998 motion picture, *Saving Private Ryan,* it falls to us to do all that we can to "earn this" — to dedicate ourselves to preserving and passing down to our heirs the legacy of human liberty and dignity that our fathers defended at such staggering cost.

As I write this, only 119,000 of the sixteen million Americans who served in WWII remain alive — less than one percent — and their number is rapidly shrinking with every day that passes. At the pace this work is proceeding, they may all be gone by the time this reaches print. I am so grateful that Pop left behind a written record of what he saw and heard during those momentous times. That way, even though he is gone, his voice still speaks. I pray that you will find that voice as compelling, inspiring and, yes, even at times as funny, as I have.

— David Wayne Miles
January 1, 2024

Birmingham, Alabama
May 12, 1972

This album was originally prepared over a period of many months ending March 25, 1950, at Savannah, Georgia. Unfortunately it was prepared on both sides of the pages of an inexpensive album. Twenty-two years of handling and aging caused the pages to crumble. I have this date undertaken the painstaking removal of the original items with knife, scissors and razor blades and re-entering them into this newer album. After only one day's experience, the formidability of the task suggests it may never be concluded successfully. The following is quoted from my original introduction:

War is only a part of history, and a small part at that, yet to the ones actively participating and their loved ones, no part of history is more seriously fraught with heartache, intense strain and extraordinary events. Because our memories are short (in many instances a heaven-sent gift), and our desire to pass on to others those happenings in our lives that will likely least interest them is so great, many have set about to record their personal experiences during World War II. Since I make no claim to being above these selfsame drives, forthwith is recorded my not-so-remarkable account.

If it seems to the reader that far too many details (dates, places, names, organizations, etc.) have been recorded, let it be remembered that this book is prepared for MY benefit.

Ralph J. Miles

Reproduced above is Ralph's own introduction to his memoirs, taken from Volume 1 of what he simply titled, *War Albums*.

Prologue

At the core, the American citizen soldiers knew the difference between right and wrong,
and they didn't want to live in a world in which wrong prevailed. So they fought, and
won, and we all of us, living and yet to be born, must be forever profoundly grateful.
— **Stephen Ambrose**

February 16, 1943

It was a typically mild mid-winter day in the languid city of Savannah, Georgia. Even more pleasant than the temperature on this particular day was the humidity — or more precisely, the lack of it. In fact, the air was down-right breathable, not heavy with suffocating moisture like it is most of the year. Appreciating this very agreeable weather was a wiry, well-dressed young man in his early twenties, who sported a healthy pile of dark, wavy hair combed straight back and parted on the left. But he was not on a leisurely stroll to enjoy the weather; he was on a mission, and he strode resolutely down crowded Broughton Street, the city's main drag, clearly intent on doing what he had left his busy desk at the Savannah FBI office to do.

As Ralph made his way through the bustling throng crowding the Broughton Street sidewalk, a photographer snapped his picture as he passed by. It was the Great Depression. Everyone was looking for a way to make an extra dime during those hard times — selling passersby their candid pictures was a way a fellow could do just that. Instinctively empathetic with the vendor's cash-poor plight, Ralph paused just long enough to dig fifty cents out of his pocket, pay for the photo and resume his way down Broughton toward his destination, the Liberty National Bank building. He had important business there, and he had no intention of being late.

He hated being late. In fact, Ralph abhorred any form of sloppiness or laziness. All he had ever known throughout his entire hard scrabble life was grinding work, done to the utmost. He had learned early in life that money was nothing more than congealed sweat. "Whatever you do, do it with all your might," was his mantra.[1] And the fact is, Ralph's deeply ingrained ethos of hard labor, exhaustive preparation

---

1. A quote from Ecclesiastes 9:10: "Whatever your hands find to do, do it with all your might." The Scriptures were Ralph's lodestar from age fourteen until the day of his death seventy-three years later.

and minute attention to detail had stood him in good stead throughout his life, and he wasn't about to change now.

He had obtained a good job soon after graduating from recently opened Savannah Commercial High[2] in 1939 as an office boy for an insurance company, earning the princely sum of fifty dollars per month.[3]

Not long thereafter, he received a call for an interview from an organization that had recently opened an office in Savannah: The Federal Bureau of Investigation. The Bureau needed a night clerk to monitor and respond to communications from 6:00 pm until 8:00 am. Unbeknownst to Ralph, the Bureau had called Savannah High to see if they had any recent graduates that they could recommend for the position. Principal Arthur J. Funk unhesitatingly responded that he knew just the man for the job, Ralph Miles. During his four years at Commercial High, Ralph had developed a strong bond with Mr. Funk, who had taken a deep personal interest in the impoverished but intense, bright student. Funk often called Ralph to his office, not for disciplinary reasons, but simply to talk — about religion, politics, any topic that sprang to mind — with this quiet, serious and well-read youth. By the time he reached the end of his senior year, Ralph — an obsessive perfectionist even at that young age — had won every top award Commercial High offered: Scholarship, Leadership and Military Science. The last award came because of Ralph's outstanding service in the ROTC. (Years later, he would recall that he joined the ROTC to have a decent set of clothes to wear to school.) Mr. Funk had a special medal made for Ralph that included all these honors, a remarkable achievement never before attained — and, as far as is known, never since — by any student in school history.

Thus armed with his mentor's enthusiastic endorsement, Ralph eagerly took on his new duties at the Bureau in July of 1940. To most people, "night clerk" does not sound like much, but in the FBI, it was and is a position of considerable importance, requiring a highly responsible person with keen record-keeping skills, good memory, and an eye for detail — perfect for the meticulous, detail-obsessive Ralph. Ralph dove into his responsibilities with same sedulous vigor that had brought him success thus far in his life. Ralph quickly found a home in the FBI, and this humble beginning would provide the springboard for a stellar thirty-plus-year career interrupted only by a two-year hiatus, the result of a call from Uncle Sam three years after joining J. Edgar Hoover's team.

It was this same modest but promising soon-to-be-G-man who entered the Liberty National Bank and, after checking the directory on the wall, stepped into the elevator and firmly pressed the button

---

2. Savannah High School, the city's flagship secondary school, at first operated out of a National Guard Armory in downtown Savannah, with spill-over classes meeting in Chatham Academy, a building now occupied by the Savannah City Board of Education. In 1937, the school moved into the imposing new campus on Washington Avenue. Its first and long-time principal, Arthur J. Funk, created a school within the larger facility called Commercial High, that focused on preparing students for the business world. Both Ralph and his future-wife, Eleanor Hobbs, attended Commercial High, though Eleanor never graduated. The last senior class graduated from the 1937 campus in the 1970's. The iconic facility currently provides additional classroom space for the Savannah School of Art and Design (SCAD).

3. The equivalent of about $960 in modern currency.

next to the number 7. When the doors slid open, he walked briskly down the hall until he reached Suite 713 — Office of Selective Service Board #13. Ralph pulled open the glass-paned door and took his place at the end of a line of other young men about his same age — and all there for the same reason: to register for the draft.[4]

It seems odd at first: Ralph, a true patriot to the core, didn't volunteer right after Pearl Harbor like a lot of other guys did. Ralph loved his country deeply and unreservedly. And he was without a doubt mentally tough. No one who knew him could ever recall Ralph showing by word or deed the slightest fear of anyone or anything. He had faced down more giants by the time he was twenty-one than most people do in a lifetime.

The fact is, Ralph had a lot of folks depending on him.

Born March 28, 1921, on a sharecropper farm between Glennville and Claxton in what was then-known as the Birdford District of Tattnall County, Georgia, Ralph and his family had little cash. Daniel Dozier Miles and his family wrested a hard-scrabble existence from the red Georgia clay, moving from one ramshackle farmhouse to another in Tattnall, Evans and Candler Counties.

In 1930, the family left the country, moving to Savannah when Ralph was nine years old. The Miles family's location may have changed, but the cycle of alcoholism (at least one of Ralph's older brothers had inherited from Dozier the addiction gene that ran through multiple generations of the Miles family tree), and poverty remained the same. But Ralph always had happy memories of his childhood despite his dire material circumstances, for, as he often later remarked, "I didn't realize how poor we were, because nobody I knew had any more than I did."

Ralph once described his "Papa" as "a tall, barrel-chested, ruddy, blue-eyed Irishman — jovial, out-going, carefree, hard-drinking and improvident." According to Ralph, Dozier "was one of those unusual drinkers who did not get mean when drinking." He never physically abused either his wife Janie or any of the kids (if anything, Ralph remembered him as an indulgent father), but he was utterly irresponsible as a husband and provider. He was in many ways, not all of them good, a larger-than-life kind of character, and Ralph loved him despite his ways. After he gave up sharecropping and moved the family to Savannah in 1930, Dozier knocked around in various odd jobs, mostly selling dry goods. He finally landed a pretty decent position as a night watchman at the International Vegetable Oil Company plant in west Savannah.

Dozier's payday ritual never varied. He dutifully collected his paycheck — and then spent the rest of the day and night in the local honky-tonks, drinking away his earnings almost to the last dime and flirting with the local barfly gals. Bleary and jolly, he would then weave his way home, stumble through the front door, sit down in front of the radio, turn it on and quickly pass out. Ralph often recalled that after a while, Dozier's loud snoring would bring Janie tiptoeing into the room to turn off the radio, ever-so-gen-

4. Congress passed a law in 1940 that called for certain men between the ages of twenty-one and forty-five to register with the Selective Service. Of those who served in WW2, about 39% were volunteers and the rest were drafted. Fathers were exempt from the draft until the law was changed in 1943 — the same year Ralph got his notice.

tly. Instantly, Dozier would awake, sputtering and snorting and demanding to know who had turned off his radio show. "He snored loudly in his sleep at night, too," Ralph remembered. "So many times, I heard Mother say, 'Dozier, wake up and turn over!'"

But Dozier's prodigal chickens finally came home to roost when Ralph was fifteen years old in April of 1936. As Dozier turned a corner while making his rounds at about 4:30 am at the plant, he was brutally waylaid by an assailant with an axe. The attacker repeatedly bludgeoned D.D.'s skull with the weapon, cleaving his head down to his shoulder and spattering the walls and floor with blood and brain matter. The savage violence of the crime strongly suggested to the police that the murder was a crime of passion. Dozier's head was so badly disfigured that the undertaker had to piece it back together as best he could from photographs the family had provided. Fleeta Hobbs Miles, Ralph's sister-in-law, later remembered that Dozier's battered head was so swollen and distorted that, in her words, "he looked like a bulldog."

Police believed they had a good idea of who was behind the crime — the jealous husband of a family member — but they were unable to get anyone to testify. And Janie Miles declined to press any charges against the suspect, most likely to avoid further family scandal and embarrassment. She had to put up with enough heartaches with Dozier's antics while he was alive; best now to just remember the good parts, let sleeping dogs lie and move on. Yet, even with all that, Ralph never spoke ill of his Papa, though he was not blind to his faults and foibles.

On the other hand, Ralph often spoke tenderly of his "longsuffering saint of a mother, who desperately tried to hold body and soul together for the family on the last dime remaining from Papa's paycheck after he drank up the rest." Clearly, Janie's (her preferred name — she hated her birth name, Mary Jane)[5] quiet, stolid and enduring nature had a lasting and profoundly formative impact on Ralph's own temperament and outlook on life.

While Ralph was mature far beyond his years in 1941, his wife, Eleanor, at seventeen was not simply young — she was also emotionally fragile, a consequence of having lived through a trauma-filled childhood from a very tender age. She had suffered sexual abuse as a child at the hands of close relatives, and she lost her stern yet doting father, Alfred Glenn Hobbs, Senior, to a workplace accident when she was only eight. On the night of April 26, 1932, a fire broke out at the National Resin, Oil and Size plant in west Savannah, where Glenn Hobbs had worked for several years as a night watchman. Hobbs joined in the battle to extinguish the blaze, hosing down a large tree that had been engulfed in the fire. After much effort, Hobbs managed to douse the angry flames — or so he thought. But the tree had continued to smolder inside and, its core burned through, exploded and crashed down onto Glenn Hobbs, crushing his chest and killing him instantly.

Such a tragedy would disrupt the equilibrium of any family, but in the case of the Hobbs clan, it was a particularly shattering, watershed moment, especially for Eleanor, not yet nine years old. Eleanor always said that that they had been spared the worst impacts of the Depression as long as "Papa" was alive,

5. As a child, Janie was teased with, "Mary Jane is a pain," so she always gave her name as "Janie."

because he always had steady work and a decent paycheck. "Are we poor?" she remembered asking her mother, Pearl Hobbs, as a tot. "We're not poor, and we're not rich, Kitten," was Pearl's reassuring reply. "We're comfortable."

But all that — and more — changed with the horrific events of that April night.

Left impoverished by her father's sudden death, Eleanor and her mother moved out of their house at 516 West 41st street, first over to Tattnall Street and later to a tiny place some ten blocks away on West 32nd Street. Eleanor's older sister Mattie and her six children then moved in with Pearl and Eleanor when Mattie's husband, Bill Morgan, lost his job as a steam fitter.

And if that weren't enough, Eleanor's other sister, Alma, frequently left her two children with Pearl for extended periods while she and her husband Bartlett traveled. Eleanor had to compete with an ever-present swarm of nieces and nephews — all about her own age — for attention, affection and life's necessities from her harried and ailing mother, who had suffered from high blood pressure for some time; Eleanor remembered that she always wore loose-fitting dresses to help her cope with Savannah's stultifying heat and humidity. Living space was at such a premium in the cramped little cottage that fourteen-year-old Eleanor had to share a bed with her mother.

The pressures of poor health, widowhood and the struggle to provide for her teen-aged daughter and her grandchildren all finally took their deadly toll on Pearl Hobbs in the wee hours of a late-February morning in 1938. Shortly before 12:30 am, Eula Pearl Lee Christie Hobbs awoke abruptly from her slumbers and sat bolt upright in the bed, clutching at her head, which suddenly throbbed with excruciating pain.

"What's the matter, Mama?" an alarmed Eleanor asked, startled awake by her mother's sudden distress.

"I don't know, Kitten," Pearl answered, using her favorite term of endearment for her menopause baby.

Pearl Hobbs then fell limply back into Eleanor's arms as the ruptured blood vessel — the source of her agony — sent a deadly torrent of blood rampaging through her brain. There hissed from Pearl's throat the dreaded "death rattle," as her final breath rushed from her body. The terrible sights and sounds of that dark night haunted Eleanor for the rest of her days. Years later as a child, I often sat, both terrified and enthralled, as Eleanor and her older sisters would tell and retell the story as they reminisced about "the old days." It still gives me the shudders.

Now orphaned and utterly destitute, fourteen-year-old Eleanor moved in with her sister Fleeta and her husband back on West 41st Street. Fleeta's spouse happened to be Ralph's older brother Osceola Curtis "O.C." Miles. Fleeta was a kind and loving sister to Eleanor, but Curtis embodied many of his father's worst qualities and none of his better ones. Curtis was a gruff, crude man and a violently abusive alcoholic who barely tolerated Eleanor's presence in his house, often demanding more money for her care from Eleanor's older brother Herbert, who paid for Eleanor's food and lodging.[6] Eleanor's "room" was

---

6. In the mid-1960's Curtis finally gave up alcohol "cold turkey," after being told by Fleeta's doctor that he would likely kill her if he kept drinking — unlike Dozier, Curtis was a mean drunk. More than once, Ralph had to go pull him off Fleeta to save her life. Curtis, a leather-

a bed placed under a stairwell, with only a curtain to provide a shy, blossoming teenage girl with some semblance of privacy. Curtis made little secret of his resentment at having this additional mouth to feed, and Eleanor often recalled cowering in her "room" listening to a drunken Curtis rant about her intrusive presence under his roof. It was not a happy time, and it left additional deep scars that would torment Eleanor off and on throughout her life. The only good thing about it was that Eleanor was again living right next door to a devout, even keeled and handsome young man who had had his eye on Eleanor for some time — Ralph Miles.

Ralph, his younger sister Mary and his other older brother Fred and his family, had moved into the Hobbs family's former house at 516 West 41st Street. Ralph's mother Janie came from a stable, religious family in Tattnall County that had produced more than one Baptist preacher. To supplement her feckless husband's unpredictable income, she took in laundry from the neighbors and gave haircuts to the boys in the neighborhood for 25 cents each. Ralph said that Janie never complained about her lot, though she certainly had every right to do so. She was perfect in Ralph's eyes, and he worshipped her. By all accounts, Janie was a gentle, kind, and wonderful lady. No doubt she, like many of her generation, had learned a harsh but vital lesson for survival in such daunting circumstances as those she had to endure: *You do what you have to do.* I often heard Ralph offer that same diffident response whenever pressed to explain how he, too, managed not just to survive but to overcome the blows life landed on him and grow into the stable, successful man he was.

Something else entered Ralph's life during this time that proved powerful and life-altering: The Lord Jesus Christ. Ralph's older sister Doris, whom Ralph adored and respected, had become a member of the church of Christ that met in a nearby converted home at the urging of her future husband, Lee Rawlings. Soon, Ralph too was baptized, as was his mother, his sister-in-law Fleeta, and his younger sister Mary. Ralph invited Eleanor to go to church with him one Sunday, and it was not long before Eleanor, too, obeyed the Gospel call and became a Christian.

Providence had thrown Eleanor a lifeline in the form of a stable, focused, and godly man such as Ralph, who had clearly absorbed his mother's gentle, quiet and spiritual nature rather than the boisterous carelessness of his father, Daniel Dozier.

In late June of 1941, when he was all of twenty years old and Eleanor had yet to reach her eighteenth birthday day, Ralph married his childhood sweetheart. By March of 1943, they were eagerly awaiting the birth of their first child, Sharron Ann, due on their second wedding anniversary.

So, there it is: a young, emotionally fragile wife, a new-born daughter, a widowed mother with little income, lots of mouths to feed, a family to help house and clothe — all weighing heavily on the shoulders of the one Miles man who had a steady job, a firm spiritual compass, an education, and a bright future. All

---

tough guy, told me years later that not a day went by that he didn't crave a drink just as badly as he did on his first day of sobriety. I can't imagine how much guts that took, but we were all very grateful for his resolve.

in all, not the kind of circumstances that would impel a devoted husband, doting father, and supportive son to voluntarily march off to the deadly uncertainties of a world war.

But, if Uncle Sam called, Ralph Miles, a true patriot, knew that he would answer that summons without hesitation. It was imprinted on his DNA and engrained in his rearing: *You do what you have to do.*

And so, he did.

This is his story.

Houses occupied by the Miles family on West 41st Street in Savannah, Georgia. The Miles' moved from 512 (left) into 516 (right) in 1933 after the Hobbs family moved away.

Right: Pocket watch Dozier Miles had on him the night he was murdered at 4:30 am on April 15, 1936. Note that the watch stopped at 4:45. The watch was found on the person of a black man a day or two after the crime, and he was arrested on suspicion of the murder. The charges were dismissed as the evidence soon clearly pointed in another direction — someone within the family. Perhaps the real culprit sold or planted the watch on the hapless "hobo," but the idea that itinerant hobos broke into the plant and waylaid Dozier stuck with Ralph. Perhaps this was the story he was told by the adults in the family to avoid revealing the true, more sordid motive behind the act — unlikely, to my mind, for surely Ralph, at age 15, was sufficiently aware to pick up on whispered conversations among scandalized family members.

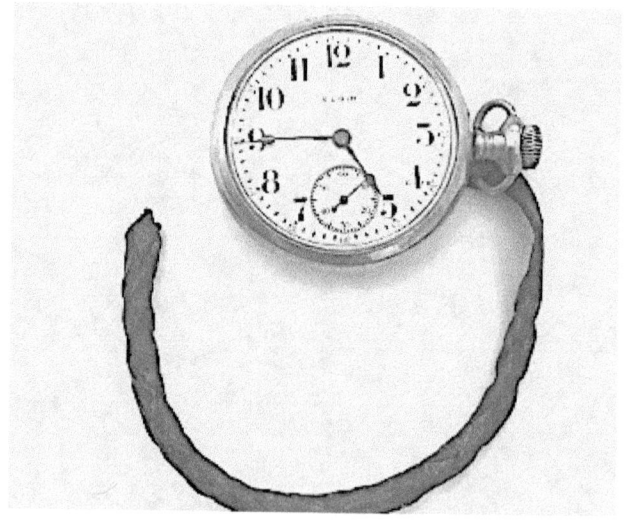

Above, from left to right: (1) Mary "Janie" Rushing Miles, Ralph's mother, on the Burkhalter sharecrop farm in Candler County, 1928; (2) Ralph, age 4, and his father (right, background), near Cobbtown, Georgia; (3) Daniel Dozier "DD" Miles, one year before his grisly murder the following spring.

Savannah (Commercial) High School awarded 3 honors to seniors: one each for Scholarship, leadership, and military science. I won all three, so the principal, Arthur J. Funk had a special gold medal made to include all three honors (left).

Left: Ralph in class at Savannah High in his ROTC uniform. **Right:** Awards Ralph won as a senior at Savannah High School, along with Ralph's own type-written description of each. In recognition of Ralph's winning every award the school offered, Mr. Funk had special medal made for his star student.

Far left: Arthur J. Funk, Principal of Savannah Commercial High School. Ralph and Mr. Funk developed a close and mutually respectful relationship during Ralph's time at the school. Ralph spoke fondly of Mr. Funk and the profound impact the educator had on the trajectory of his life.

Left: Ralph walking down Broughton Street, the commercial center of Savannah, Georgia in 1941.

Left: Alfred Glenn Hobbs, Sr. 1917. **Middle:** Pearl Hobbs in 1937, one year before her death in the arms of her fourteen-year-old daughter Mary Eleanor (**right**).

Above: April 28, 1932. Family of Alfred "Glenn" Hobbs, Senior, Eleanor's father, at his graveside in Bonaventure Cemetery in Savannah. Standing at the far left is Glenn's widow, Eula "Pearl" Lee Christie Hobbs. To her right are her seven children, standing in order from oldest to youngest: Ethel, Mattie, Alma, Herbert, Fleeta, Alfred Glenn Hobbs, Jr ("Bobby") and Ralph's future wife, Mary Eleanor.

Left: Converted home that served as the meeting place for the church of Christ where Ralph and Eleanor and several of their family were members, and where Ralph and Eleanor were married on June 29, 1941. The sign over the door, "The Church of Christ," was hand-painted by Ralph.

Left: Ralph and Eleanor on their wedding day, June 29, 1941. **Right:** Eleanor, Ralph and their daughter, Sharron Ann, who was born June 29, 1943, on their second anniversary — three months before Ralph was drafted.

1. 'This Is the Army, Mr. Jones'

As the United States prepared for war, military leaders had a long list of needs—guns, tanks, ships, and equipment of every kind. One of the things they needed most of all, however, was people. [1]

It all started with the inevitable registration with the Selective Service Board. Congress had passed a new law in 1943 that ended the exemption that had been in place for married men with children. Accordingly, I made my way post-haste to the Selective Service Office in downtown Savannah and addressed that little bit of unpleasant business in February.

Events hastened along with the arrival in August of my notice of Classification: 1-A — eligible for prompt induction! And in a matter of two short weeks arrived the most ominous document of all:

ORDER TO REPORT FOR INDUCTION
The President of the United States
To: Ralph Jacob Miles

GREETING,
    Having submitted yourself to a local board composed of your neighbors for the purpose of determining your availability for training and service in the land or naval forces of the United States, you are hereby notified that you have now been selected for training and service therein.

    You will, therefore, report to the local board named above at 713 Liberty Bank Building at 3 pm on the 4th day of Sept, 1943 for instructions prior to your departure for induction on Sept 7th.

"Greeting" indeed!

Thoroughly intimidated by that last-quoted hunk of correspondence, I hesitated not to present myself to the Local Board in the old Liberty National Bank building on Broughton Street — a Savannah landmark, sadly since demolished. On September 7, 1943, a busload of probable inductees, including me,

---

1. https://www.nationalww2museum.org/war/articles/training-american-gi

departed Savannah for Fort McPherson.[2] On the following day, we underwent a physical examination to determine fitness for service. My most vivid recollections at this station were three in number:

1. Countless numbers and numberless types of naked male figures followed one another through endless mazes of rooms, corridors, doctors, male nurses, and baleful apparatuses; gone was all modesty, all admiration for the male figure — everything gone but a longing to be outside and free from the stench of a multitude of sweating humans! The inspectors were impersonal and unimpressible, hearing impatiently (and dismissing even more impatiently) a great variety of sad tales of broken health. I often heard that sometimes the Army took in men physically unfit, but after this experience, I wondered how in the world it could not be possible. Yep, I was found quite healthy. Pass on, brother.

2. Making a choice of services (Army, Navy, Marines, Air Force). The Air Force was out as the medical examiner mumbled something about slight color-blindness. Of the remaining three, I naturally chose the Army (Because they feed the best chow? Because they have the best-looking uniforms?). Because the Army would allow me twenty-one days before reporting for active duty, which was more than any of the others did! And, for some wholly unaccountable reason, I thought I wanted to be a "soldier."

3. Signing papers, papers, and more papers.

Back home in Savannah, I made some slight pretense of setting my affairs in order, but I'm afraid it was left up to my poor wife Eleanor to handle most of the details as I was too busy trying to pack lots of memories into twenty-one days. Our only child was by now less than three months old. The house we had rented for two years at 1127 East 39th Street would have to be left as the $50[3] then paid a private in the US Army would not cover the rent and other expenses. The furniture would have to be stored.

Reluctantly and with grave misgivings I said goodbye to my loved ones and reported to Fort McPherson, for active-duty September 29, 1943, Army Serial Number 34 828 227. I began a diary of sorts, a common affliction among new recruits; it came in handy years later, helping to jog my memory as I put together my memoirs of my time on the Front. Busy days followed: issuance of ill-fitting uniforms, issuance of pay record, "Dog Tags," innumerable readings of the Articles of War, viewing many stom-

---

2. Ft McPherson, named for Union Civil War General James B. McPherson, opened in 1885. The facility processed thousands of inductees into the armed services during WWII. The fort was also the site of an internment camp for Japanese Americans. Later during the War, it served as a discharge center for servicemen who had completed their tour of duty. It was deactivated in 2011 (https://www.georgiaencyclopedia.org/articles/government-politics/fort-mcpherson/).

3. About $865 in 2023 dollars.

ach-churning venereal disease movies, taking various shots[4] and taking tests. The Army gave us three tests: General Classification (my score — 119), Mechanical aptitude (score — 121), and Radio aptitude (my highest score — 125).

After a long and exhausting day of being "processed," I dropped into my bunk and quickly went sound asleep.

AT 5 AM ON OCTOBER 4, a heavy hand shook me from my slumbers, and I was on my way to my first real camp. Destination unknown (situation normal) but to my surprise, a short trip to Anniston, Alabama, home of Ft. McClellan — an Infantry Replacement Training Center at that time.[5]

For the first two days I was in a rough, tough company destined solely as front-line infantry replacements. The officers treated the men like the morons many of them were[6] and I sharply recall the toughest sergeant I've ever seen (before or since) telling us what woe would befall us if any of us used the wash basins for urinals. The reason for such unusual cautioning became apparent when I first entered the latrine room — the difference between the wash basins and the urinals was scarcely discernible!

Luckily, I was transferred to Company C, 5th Regiment, 15th Battalion on October 6, 1943. This was a company of prospective communications replacements. The officers were more relaxed and the men patently more intelligent. Here I learned that because of the high score I made on the Radio Aptitude Test, I would have the opportunity to learn some minimal data about the Army's radios — that is, if I was able to live through the initial basic training required for all recruits. Basic training at McClellan was, as is all basic training, tough; but good for me physically. After four months of battling with foot blisters, Alabama clay, night hikes of ten miles, dry firing, bayonet practice, and daily daylight calisthenics, I weighed in at 160 pounds (up from 145 when inducted) and was in the peak of physical condition. On the rifle

---

4. Flu, typhus, dysentery, syphilis — plus botulinum toxoid for protection against botulin toxins often used as part of biological warfare. The Allies had received what turned out to be faulty intelligence that the Germans had loaded botulin toxins onto their B-1 "buzz bomb" weapons.

5. First opened in in 1898 under the name "Camp Shipp," the fort was re-named after Civil War Union General George B. McClellan during WWI. With the outbreak of WWII, the Camp expanded dramatically and was re-designated a Fort. In 1943, Fort McClellan became, as Ralph notes, a major Infantry Replacement Training Center (IRTC). After nine weeks of basic training, the IR's ("infantry replacements") were sent to other Army installations for further instruction either in combat or in a specialized service area, such as radio operator. Also at Ft. McClellan was a POW internment camp as well as a cemetery for German and Italian POW's who died in captivity at various POW facilities in Alabama and the Southeast. The Fort was deactivated in 1999. (Geni Certain, "Fort McClellan," http://encyclopediaofalabama.org/article/h-3525).

6. Lest anyone think Ralph was exaggerating, "Psychological examinations for draftees regarded as below normal in intelligence were adopted 1 August 1942. Restrictions on the number of illiterates accepted were … removed … 1 June 1943." (The Personnel Replacement System in the United States Army, by Lt. Col. Leonard Lerwill, Department of the Army, 1954, p. 313).

range, I scored 172, which ranked me mid-way between Sharpshooter (165) and Expert (180). My most vivid recollection: exercising outside in fifteen-degree temperature, stripped to the waist!

Incidental to basic training, we learned to send and receive International Morse Code to a relatively proficient degree.

Living quarters at Ft. McClellan were anything but ideal. We were housed in temporary buildings called hutments or huts for short. The winter winds shrieked through the thin walls and flapping doors of our huts at will with nothing but the curses of intemperate men to stop them. When I was assigned to the Birmingham FBI office some twenty-five years later, I tried to locate my area and show my family where I lived during basic. I asked a young MP at the gate for the location, and he instantly allowed that he had no idea as he "wasn't even born twenty-five years ago!" I found out later that the huts had all been torn down.

This cartoon (**below**) is an exact reproduction of the corner of Hut #9, with an **"XX"** (inside black circle) marking my bunk. (Scratched out from the original cartoon during the process of removing it from my first album, is Kleman's face peering out from inside the coal-fired stove at the Sarge).

"Come, come, Kleman, and what are you going to do when it really gets chilly?"

There were some fine men in my hut, some of whom I would be proud to name among my permanent friends: Michiganders Paul G. Hamilink, John Nivala and Howard Mathison, E.J. Nijoka (an Illini), and fellow deep-Southerner J.C. Stuckey from Mississippi. But for my blushing modesty, I would admit that I was named Hut Commander, my "badge of rank" being a black armband.

Lest I forget, I was also named leader of Squad 1, 3rd Platoon. Wait ... now that I think about it more, I believe that the armband denoted squad leader rather than hut commander! My high school ROTC gave me the edge for this exalted position!

In December of 1943, Eleanor and Sharron Ann — born on our second anniversary — came to Anniston to join me. Previously, in Savannah, we had become acquainted with a couple who were originally from Anniston — Jerre and Lou Watson. In Anniston we found ready friends in their parents, Dr. Jerre and Mrs. Annie Watson and their daughter Rebecca. These very fine people helped us find quarters for Eleanor and Sharron with Mr. and Mrs. C.A. Voss[7] on Blue Mountain, an area of town on the Ft. McClellan side of Anniston. We enjoyed the Voss family. Also rooming with the Voss' were Ruth Feazell and her daughter Rita.[8] Mr. Feazell was away in the Navy. Eleanor and Ruth became good friends — a very good thing for my poor bride, having been snatched away from hearth and home to this frozen foreign land called Alabama.

The radio part of my training at Fort McClellan ended in early February of 1944. I got a card for my efforts attesting to my new-found status as "radio operator — SSN 776." As best I recall, "SSN 776" referred to the military Service Specialization Number, or similar wording. The ones achieving the highest radio scores were selected for additional training. I was one of the fortunate ones from my group and on February 11, 1944, I found myself at The Infantry School at Fort Benning in Columbus, Georgia.

While official life at the Infantry School seemed plush (brick barracks, hot water, decent food, good beds with clean sheets, passes to town at any time not on duty) after the rigors of Fort McClellan's "huts," the reception of my family in Columbus was practically indecent. Columbus' inhospitable attitude toward soldiers is explained by the fact that Fort Benning was the largest Army post in the country — and the locals had become as sick of the constant flow of transient soldiers (of widely varying degrees of moral and behavioral rectitude) as the soldiers had become of the cold and resentful town residents.[9] We survived, however, without apparent permanent damage.

---

7. Charlie A. Voss was a veteran of World War I, which no doubt contributed to his sympathetic and welcoming disposition toward Ralph and his little family. Charlie was part of the 2nd Casual Company in the First World War – in WWI "casual" units consisted of soldiers who were between (or awaiting) permanent assignment, during which time they performed various critical tasks not covered by permanent units, such as transportation duties. The "official" DOD definition of "casual" is "transient" (https://www.militaryfactory. com/dictionary/military-terms-defined.php?term_id=857). Charlie passed away not long after the war in 1949. His good wife survived him by nearly thirty years, passing in 1976.

8. Ruth's husband was James Edward Feazell. He, like Ralph, was a radio guy, an Aviation Radioman Petty Officer 3rd Class, or ARM3 in Naval jargon. James survived the war and passed away in 1985. The Feazell's apparently were natives of Anniston, as both they and their little girl, Rita – who, sadly, died in 1946 — are all interred in Edgemont Cemetery in Anniston.

9. Phoenix City, a town directly across the Chattahoochee River from Columbus in Alabama was a notorious hotbed of prostitution,

We lived in a one-room apartment at 1018 15th Street that we rented from a Mrs. Robinson, who lived in a house behind us. But for the friendship and companionship of fellow-soldier Willis Green and his wife Lucille, the one-room hut we occupied at the then-outrageous sum of $30[10] per month would have been unbearable. Willis and Lucille rented a single room in Mrs. Robinson's home.

It was in the first of two courses at The Infantry School, "Enlisted Communications Course, Class #42," that I became more intimately acquainted with both the operation and maintenance of infantry radio sets.

In May I finally graduated from the Enlisted Communications Course. But this was merely the signal for beginning another course, the "Enlisted Radio Repairman Course (ERRC)."

Those chosen to attend the ERRC found in Class #7 some men with a refreshing sense of humor. The *ERRC's Bible* was the brainchild of two of the inmates — pardon me — two of the students: Sgt. Robert Lee and Private Walter Gorg. It accurately conveys with grim humor the mysterious intricacies of radio instruction to us novices.

---

gambling, political corruption and myriad other crimes and social ills that often existed in and around "Army towns." During the war, George Patton "threatened to roll his tanks across the river...and destroy it." In 1954, Albert Patterson was nominated as the Democratic candidate for state Attorney General of Alabama, running on a platform of cleaning up Phoenix City, but he was assassinated outside his law office during the campaign by members of the Phoenix City criminal machine. Patterson's son took up his father's crusade and, upon winning the election, cleaned up the city for good with the help of General Jack Hanna of the Alabama National Guard.(Max Shores, "Up From the Ashes: The Rebirth of Phenix City" — https://maxshores.com/up-from-the-ashes-the-rebirth-of-phenix-city/)

10. $479 in today's currency.

Above: Yep, that's the way we all felt after the first few days of delving into the bewildering world of radio maintenance and repair. Somehow, we all got through it — Army must have needed radio operators, else I don't how anyone passed.

Below: Sample pages of Lee and Gorg's "Bible."

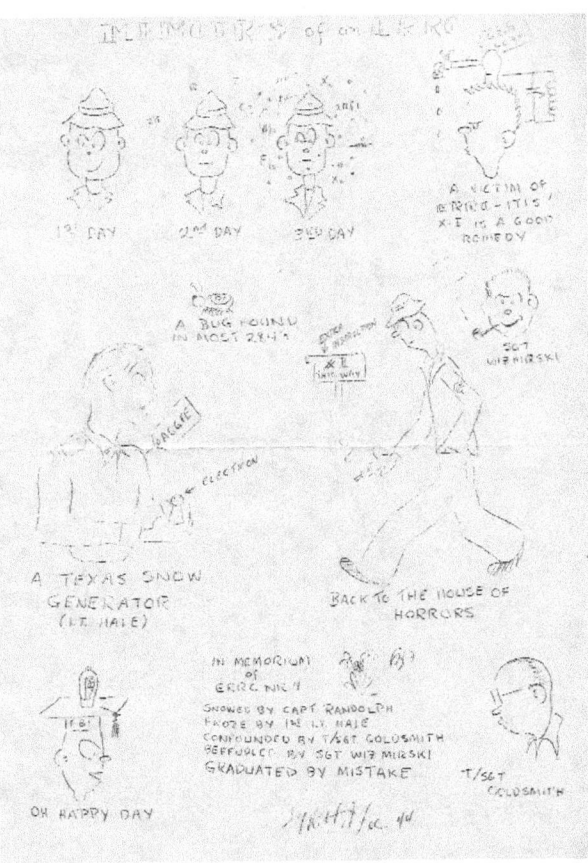

The Infantry School
United States Army

━━━◆━━━

This is to certify that

Private Ralph J. Miles
34428227

has successfully completed the

Enlisted Radio Repairman Course

given during the period

27 May 1944

to

11 July 1944

For the Commandant

Philip H. Kron

Colonel of Infantry
Secretary

In this context, "successfully completed" means only that I survived.

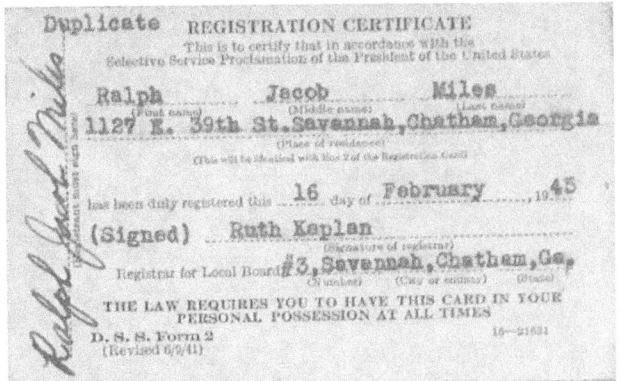

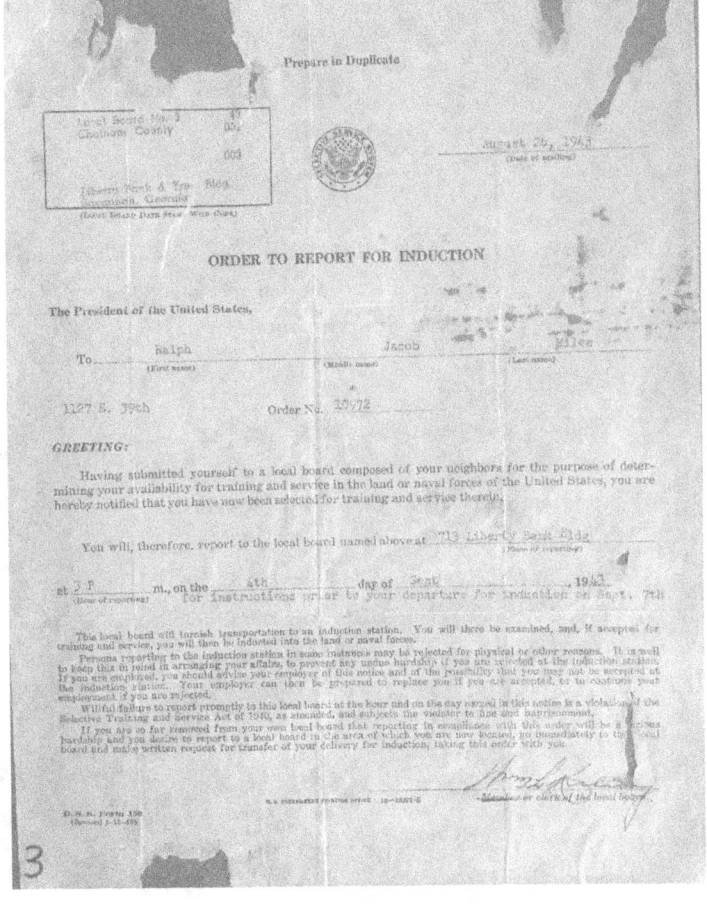

Clockwise from top left: (1) Notice of classification: 1-A. (2) My draft registration card confirming that I had in fact registered for the draft; (3) My daunting letter from Uncle Sam "suggesting" that I report forthwith for duty.

Name Ralph Jacob Miles

Army serial No. 34828227

Grade pvt.

Years of service none
(On date of opening this book)

Insurance, amount and class $ 10,000 N

Insurance premium, monthly $ 6.60 N

Allotments, amount and class $ 21.00

Compulsory allotments, amount and class $ 22.00 F

Pay reservation, class A $

Technician grade

Additional pay for FS-M AUG 1944

Person to be notified in case of emergency:

Mary Hobbs Miles USN
(Name)

wife
(Relationship; if friend, so state)

1127 E. 39th St. 516 W. 41st
(Number and street or rural route; if none, so state)

Savannah 517 Ga.
(City, town, or post office)   (State or country)

Date of opening this book Sept. 9, 1943

Ralph J. Miles
(Signature of enlisted man. Name, grade, and arm or service only. Do not enter organization) army

**SOLDIER'S INDIVIDUAL
PAY RECORD**

Above: My pay record, issued at Ft McPherson. Eleanor and Sharon Ann had moved into 516 West 41st Street in Savannah with my mother and younger sister Mary (whose husband was also in the military) and her two children to save money.

Right: For two-plus years, these "dog tags" hung undeviatingly from about my neck. All servicemen's tags bore the same information, in order: Last name, first name, MI, serial number, years of tetanus shots [T43 44 = 1943, 1944], Blood type (mine = A), and religious preference ("P" at the bottom right, for "Protestant").

Above: January 10, 1944 — Hut #9, Company C, Squad 1, 3rd Platoon, 15th Bn, 5th Regt, Infantry Replacement Center, Ft. McClellan, Anniston, Alabama. **Squatting, L-R:** Carpenter, Dolly, Cohen, MacKnight, ?, Carlson, Goodwin, Mallet. **Standing, L-R:** Van Laarhoven, ?, Arnold, J.C. Stuckey, Hosford, Paul Hamelink, Cowen, Miles. My black armband denoted my position as squad leader. On a side note, the building behind us in the photo is the most important structure on the base other than the mess hall — the latrine.

Above: Me standing in front of Hut #9, Ft. McClellan, Alabama, January 10, 1944

The humorous card pictured above may be better explained by stating that the letters "QLO" in radio language mean "use"!

Below: The radio part of my training ended February 5, 1944. I got a card for my efforts, attesting to my new-found status as "radio operator — SSN 776." As best I recall, "SSN 776" referred to the military Service Specialization Number.

CERTIFICATE

This is to certify that

Ralph J. Miles   ASN 34828227

has satisfactorily completed, during the period

from 6 Dec 1943 to 5 Feb 1944

the schedule of specialized training prescribed for a

Radio Operator   SSN 776

William J Voelker
William J Voelker, 1st Lt, Inf
Senior Instructor   Radio   School

John L. Brown
Major, Infantry
Commanding 15th Battalion

Ground portable radio set installed in jeep. Transmitter operated by telegraph key attached to the leg, or by voice telephone.

Radio men operating the old Signal Corps Radio (SCR) — 284. Picture at top right shows the SCR-284 complete with "torture grinder" (otherwise known as the generator) for field use where no vehicle power source is available, which is the way I would later become more intimately familiar with it.

Ralph's hand-written roster of Replacement Company "C", 1st Battalion at Ft. McClellan, Anniston, Alabama in February of 1944. Ralph's scrupulous attention to detail, accuracy and record-keeping that helped him succeed in life to this point continued to serve him (and his descendants and historians) well!

Left: Ralph holds Sharron in front of the house on Blue Mountain where Eleanor and Sharron lived while Ralph underwent basic training at Fort McClellan. **Right:** Ralph and Eleanor pose in the front yard of their rented home during their time in Columbus, Georgia. The house no longer stands.

Above: World War II-era Army inductee reception center at Fort McPherson in Atlanta, Georgia. Ralph mustered in here from 29 September to 4 October 1943.

Left: Willis Green and Ralph, taken on Easter Sunday, April 9, 1944, in Columbus, Georgia.

Top image: Buckner Hall, the Headquarters Building at Fort McClellan in Anniston, as it appeared during World War II. **Bottom image:** This modern image of Buckner Hall reflects little change in the building's appearance. The houses located around the circle in front of Buckner, once occupied by Army officers, are now private residences. Situated behind and on either side of the Hall are enlisted men's barracks, which, though unoccupied since decommissioning in 1999, remain in relatively good condition.

On October 9, 2021, David Miles visited the site of the house where Ralph, Eleanor and Sharron
Miles lived in Columbus while Ralph underwent radio training at Ft. Bragg.

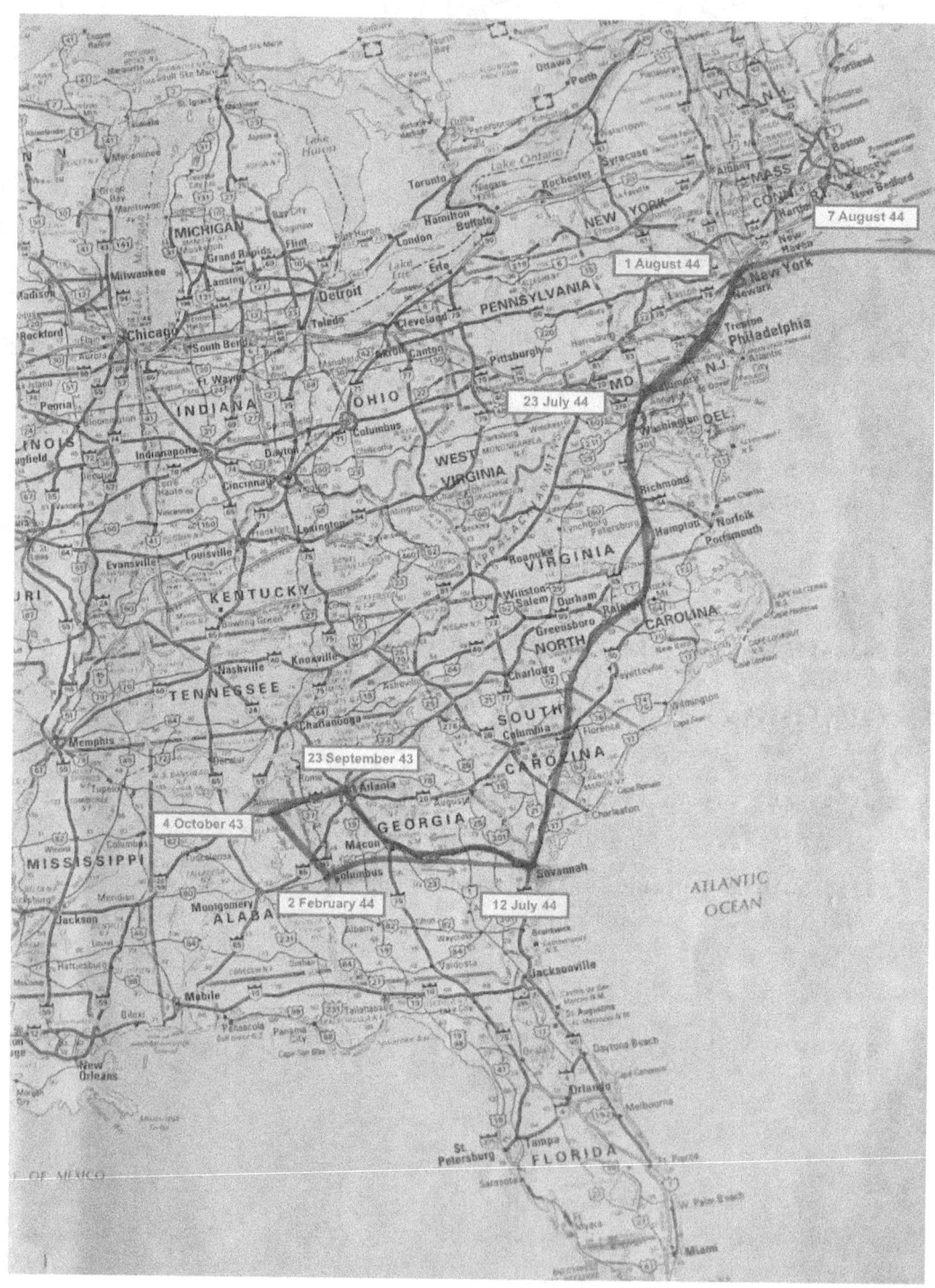

Map on which I traced my movements stateside from the time of my induction at Ft. McPherson on September 23, 1944, until my departure for England from Camp Shanks, New York on August 7, 1944.

2. Saying Goodbye

The true soldier fights not because he hates what is in front of
him, but because he loves what is behind him.
— **G.K. Chesterton**

On July 11, after graduating from The Infantry School at Fort Benning, I received ten days furlough with travel time before having to report to a port of embarkation. This was a sad furlough for we all knew it would be a long time before we would see each other again (we refused to face the alternative). Mother was terribly sick with cancer even at this time and in my heart, I feared I would never see her again. Also, although unknown to either of us at this time, Eleanor became pregnant with our first son, Ralph Junior, during this last brief time together.

The day we both dreaded, July 23, 1944, came far too quickly — our last day together before I left for Ft. Meade, Maryland. Before tearing myself away from my family, we snapped some pictures in front of my childhood home — 516 West 41st Street. Our smiles belied the fear and anxiety we felt as the reality of what lie ahead kept pushing itself rudely into our thoughts. Eleanor and Sharon would continue to live at 516 with Mother and my sister Mary and her two children while I was overseas.

I left home on the afternoon of the following day, arriving at Fort George G. Meade, Maryland on July 24th. I was assigned to Company B, 1st Battalion, 1st Replacement Depot #1, which I supposed to be my last station before leaving for overseas.[1] My memories of this post are already hazy — although I distinctly remember rolling out all my equipment in the sand for inspections and changes so many times that I finally carried away more sand than equipment! Here I signed over to my wife power of attorney, made

---

1. Ft. George Gordon Meade, still a large and very active military base, opened in 1917. In 1943, Ft. Meade was chosen to house the Army Ground Forces Depot Center Number One. At the War's outset, Ft. Meade served as a training center for new recruits and as a prison camp for Axis POW's as well as immigrant descendants of Axis nations convicted of espionage activities. At the end of the War, it processed nearly half a million returning soldiers into civilian life. The fort is located just east of Interstate 95 some thirty miles northeast of Washington, D.C., in Maryland. (An Illustrated History of Fort George Gordon Meade, pamphlet published by Fort Meade Museum — retrieved from http://www.314th.org/camp-meade/Fort-Meade-Camp-Meade-Illustrated-History.pdf).

my will, applied for overseas pay and an additional allotment from my pay for Eleanor. Special emphasis was placed upon the desertion statutes[2] now that we were about to embark.

While I was at Fort Meade, I went into Arlington to visit my sister-in-law Mattie Morgan[3] and her family and to go to church. I also called Eleanor from Mattie's house. This was the last time I was to hear her voice for more than a year.

On the night of July 31, 1944, we packed our gear and entrained for Camp Shanks[4], New York, arriving about noon August 1, after traveling all night. Even now I've forgotten all that the letters in my new address stood for, but for the record they were Co. D, 4th Pl., Shipment GN 900a, APO #15413, Inf. D-4 c/o PM, New York, NY. The most striking thing was the huge mess halls under whose steaming tops literally thousands of men were fed simultaneously. A man could eat as much as he wanted and might gain several pounds at a sitting — but he lost more than that in perspiration! After several promises of a trip into New York City which never materialized, we were finally readied for shipment on August 6, 1944.

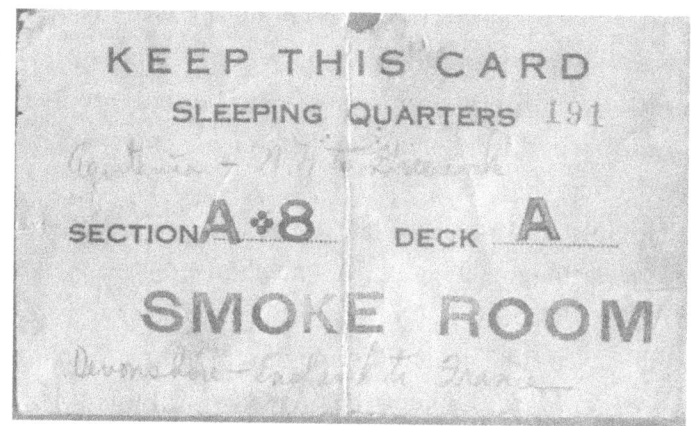

2. More than 21,000 GIs were convicted of desertion during WW2, with 49 of those sentenced to death. However, only one American soldier, Private Eddie Slovik, an Infantry Replacement assigned to the 28th Infantry Division in France, was ever executed for desertion during the war. In fact, Slovik is the only American soldier put to death for leaving his post under combat conditions since the Civil War (https://en.wikipedia.org/wiki/Desertion).

3. Mattie Marcela Mae Hobbs Morgan, Eleanor's second-oldest sister, had married William T. "Bill" Morgan. They and their children had moved from Savannah to Arlington, Virginia, for Bill to find work during the Depression. They ended up staying there for the rest of their lives.

4. Camp Shanks, named in honor of Major General David Shanks, a notable U.S. Army officer from the Spanish-American War, the Philippines and WWI, is unusual in that its history spanned only the WWII era. Despite its short lifespan, the camp served a critical role in the war effort, being the largest port of embarkation for troops headed for the ETO — 75% of the D-Day assault force passed through its gates en route to England and Normandy, thereby earning for itself the title, "Last Stop USA." The Camp closed in 1946 and only a marker and a small museum remain to mark its location. (https://en.wikipedia.org/wiki/Camp_Shanks). (https://www.historynet.com/camp-shanks-last-stop-u-s-a/)

We were rushed out to the docks early in the morning with mountains of equipment, wearing wool uniforms and OVERCOATS (temperature — 90 degrees in the shade!). It was late afternoon, however, before we were allowed to go aboard the huge British liner, the *HMS Aquitania.* As we boarded, we were handed cards indicating our quarters aboard ship (**previous page**). Of course, we couldn't have found the quarters without help since the ship was almost like a small city within itself. Loading aides herded us down the aisles and squeezed us into our four-decker bunks with all our equipment (overcoats on) to clear the traffic lanes until the ship was loaded.

There we lay for what seemed like ten or twelve hours but probably was only about three hours. Sweat oozed out of my overcoat and puddled in the canvas bunk. I nearly had a stroke it was so hot. Finally, the "all clear" signal came, and we all stripped to our birthday suits, dipped out our bunks with our clothes, wrung them out in the lavatory, and (like the scared, worried and anxious guys we were) climbed back into the bunks to await the departure. It was a long night but finally sleep came.

THE WHITE HOUSE
WASHINGTON

TO MEMBERS OF THE UNITED STATES ARMY EXPEDITIONARY
FORCES:

You are a soldier of the United States Army.

You have embarked for distant places where the war is being fought.

Upon the outcome depends the freedom of your lives: the freedom of the lives of those you love— your fellow-citizens—your people.

Never were the enemies of freedom more tyrannical, more arrogant, more brutal.

Yours is a God-fearing, proud, courageous people, which, throughout its history, has put its freedom under God before all other purposes.

We who stay at home have our duties to perform—duties owed in many parts to you. You will be supported by the whole force and power of this Nation. The victory you win will be a victory of all the people—common to them all.

You bear with you the hope, the confidence, the gratitude and the prayers of your family, your fellow-citizens, and your President—

Franklin D. Roosevelt

The next morning when we awoke, we were at sea. Red Cross workers came around and passed out trinkets and a letter from the President. This was my first sea voyage and believe me I was grateful for the greatness of the ship. To you landlubbers, this means that the greater the ship, fewer waves are able to rock it and, of course, fewer sick passengers.

It grew quite chilly at sea and after a while downright cold. Speculation was rife that we were headed for the northern islands — Greenland, Norway, etc. Naturally, no one bothered to tell us any facts. The ship was completely without escort, which fact came as an unpleasant surprise to us all until a cooperative seaman explained that the *Aquitania* was so fast that no sub could keep up with it. He also called our attention to the hitherto unnoticed fact that we were constantly changing course, running in zigzags. The sailor said this was to prevent submarines from lining up torpedo sights on us. To us novices, this was small comfort. As the days passed, though, we became more and more satisfied to just let the navy take us on over without our help or advice.

Eating on shipboard was vexing problem. There were two feedings a day which lasted from sun-up to sun-down. As soon as we finished one meal, we got in line for the next. I have no good thing to say about the chow. But as the ship slipped smoothly along, I kept wondering if I hadn't made the wrong choice back there at the reception center at Fort McPherson!

Almost the entire shipload of about 8,000 men consisted of technical replacements — cooks, bakers, mechanics, radio operators and repairmen, engineers, clerks and the like, and I was very hopeful of landing a good job — say somewhere nice like London or Paris. This uncalled-for assumption rendered the trip almost pleasant.

During the early morning of August 14, 1944, we were awakened by a commotion aboard ship greater than usual, and made our way to the upper decks, there to be greeted by a most inspiring sight. The sun was just coming up and we were sitting perfectly still in a beautiful harbor surrounded on every side by the bright hillsides of Scotland! I don't know whether my wonder at the idyllic landscape was inspired more by its natural beauty or by the fact that I hadn't seen land for a week! As our ship slid through the waves toward its berth, I managed to enhance my reputation among my fellow passengers to some extent by reading the coded light flashes between the ships in the harbor. It was one of the few times my radio "larnin'" ever stood me in good stead.

Before many hours we unloaded on the docks of the little town of Greenock, Scotland, which is near Glasgow.[5] Soon we were entrained for England proper, southward. And those trains! Wooden construction, extremely small, like toy trains. Very efficient, however — crowded, too, just a bit — I found a pic-

---

5. It is understandable that Ralph believed that he and his fellow replacements had landed at Greenock. However, they actually disembarked at Gourock, Scotland, a seaport bordering the western edge of Greenock. In fact, the two cities constitute a single continuous metropolitan area. Gourock was a primary port of entry and departure for US troops arriving at and shipping from the European Theater during the war.

ture in *Stars and Stripes* that provided a graphic illustration of how we traveled the length of the country, north to south.

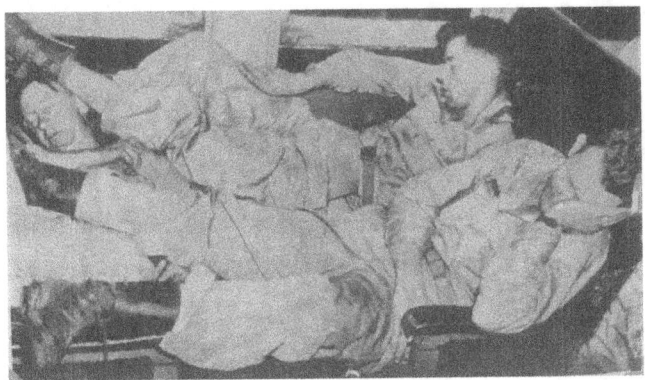

The Scottish and British countryside was a thing to be marveled at after the comparative wilds of America. Everything seemed to be in miniature. Roads were about one-half the size of ours. The most unusual thing was the absolute lack of anything except stone and brick buildings — no wooden buildings of any kind. Even fences were neat little rock rows or rows of hedge. Every acre of ground in the country is cultivated right up to the edge of the road. There is no such thing as "brush" wasteland. Being a low lander myself, the gently rolling hills were a sight within themselves. Fields were neat as a piece of paper; cow pastures looked like billiard tables. I longed for a camera to record the beauty of the country.

In this fashion (**above**) we arrived at Wellington in Southern England on Tuesday, August 15, 1944, where we were issued an Army PX (Post Exchange) rations card (**below**), which we didn't get to use to any great extent. "ETO" meant European Theater of Operations, but this card was good only in England. At Wellington we were stripped of our overcoats (now that we needed them — it was cool in England at night) and exchanged almost every item of equipment several times, ending up with approximately the same equipment we had to begin with.

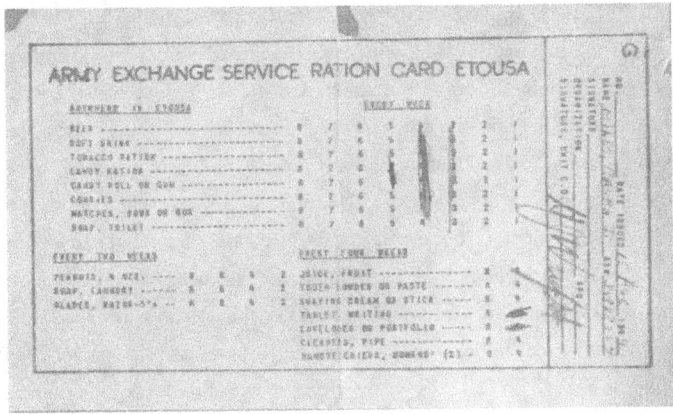

On August 17, 1944, a Lieutenant Colonel corralled us all and made a little speech. I'll paraphrase:

"Now listen to me, you guys. All of yuh came over as "specialists." Well, we don't need specialists right now. What we need are men for the line. Looks like you're it."

I could see all my careful training going up in smoke. But by this time, we'd learned to take things pretty well as they came, so without so much as a second gasp, we sauntered off to discuss the new prospects. In a few minutes, some trucks arrived, and several names were read off — mine included. We hopped aboard and were driven to Stapley, England[6], not many miles away. Here we again exchanged equipment with supply sergeants several times, eventually ending up with about the same as at the beginning. We had every item of clothing and equipment stamped: "Co. EE, Pkg. "X"-70-G, 204th Co.," whatever that meant. We didn't much care by this time.[7]

---

6. Camp Stapley was an American troop transit camp where incoming US troops were housed and then dispersed to wherever they were needed in the ETO. Located nearby at Culmhead was an RAF base. On D-Day, the 82nd and 101st Airborne Divisions were flown to Normandy beaches from a base at Upottery, a few miles southeast of Camp Stapley. The troops lived in shelters called Nissen huts. For an excellent first-hand account of life at Camp Stapley, including Nissen hut pictures, see "Story," by Robert F. Gallagher, Chapter 12 — Camp Stapley, England."

7. The Infantry Replacement System in place during World War II ultimately proved to be deeply flawed. One glaring issue was the failure in many of the IR Depots — known as "Repple Depples" in GI slang — to properly equip the IR's as they awaited their "call" to the front lines. In the words of one Army analyst writing after the war, "many replacements arriving from the United States carried weapons other than those which they should have received under the tables of organization and equipment … this caused much unnecessary work and readjustment … Troops were sometimes sent forward overburdened with unnecessary personal property but not fully equipped with required articles" (Lerwill, p. 458). In addition to poor training and equipment, the troops languishing for weeks or even months in these camps endured boredom, angst and low morale. (https://wikivisually.com/wiki/Replacement_depot). It is little wonder the front-line doughs received these poorly prepared replacements into their ranks with suspicion, skepticism and aloofness (Eric William Klinek, *The Army's Orphans: The United States Army Replacement System in the European Campaign, 1944 — 1945*, p. 136, referred to hereinafter as "Klinek"). Ralph's account of his own experiences as an IR awaiting assignment closely echoes the recollections of other IR's, as well as post-war research on the replacement system by military analysts. Two of the best such analyses are *The Personnel Replacement System in the United States Army*, by Leonard L. Lerwill Lt. Col, U.S. Army (Department of the Army pamphlet, 1954) (cited hereinafter as "Lerwill"), and the above-cited work by Eric Klinek.

Those who know me well realize that I have not the patience nor the inclination to collect stamps, so I hasten to explain that the stamps adjacent hereto, as well as those which follow at the end of this memoir, were picked up (that's a nice way to put it) from someone who *used* to collect stamps in England and Germany. The coins above are worth in the neighborhood of two dollars altogether. If they are now missing, it will mean (1) that someone stole them, or (2) I got desperate and exchanged them for American money.

Left: Photos taken on that final precious day Ralph spent with Eleanor and Sharron Ann before shipping out — brave smiles hid aching, worried hearts.

These pictures were taken July 23, 1944, in front of the same house where Ralph grew up, and where his mother, Janie, still lived — and where she would die — 516 West 41st Street, Savannah, Georgia.

Unbeknownst to either Ralph or Eleanor, Eleanor was pregnant with their son Ralph, Junior, who would be born while Ralph was pursuing retreating Nazis in Bavaria in April of 1945.

Above: *HMS Aquitania* delivering US troops to the port of Gourock, Scotland. The picture source does not date the photo, but it does note that 8,000 US troops were on board — the same number that sailed with Ralph on this same ship in August of 1944. (Source: https://hvmag.com/life-style/history/remembering-camp-shanks/

On the map **below**, the notation in the white square is from Ralph's daily diary that he kept from the time he mustered in in 1943 until his departure from Le Havre, France, for the States in July of 1945. Also on the map is a photo of the kind of housing used by troops during their time in Stapley, England: "Nissen huts."

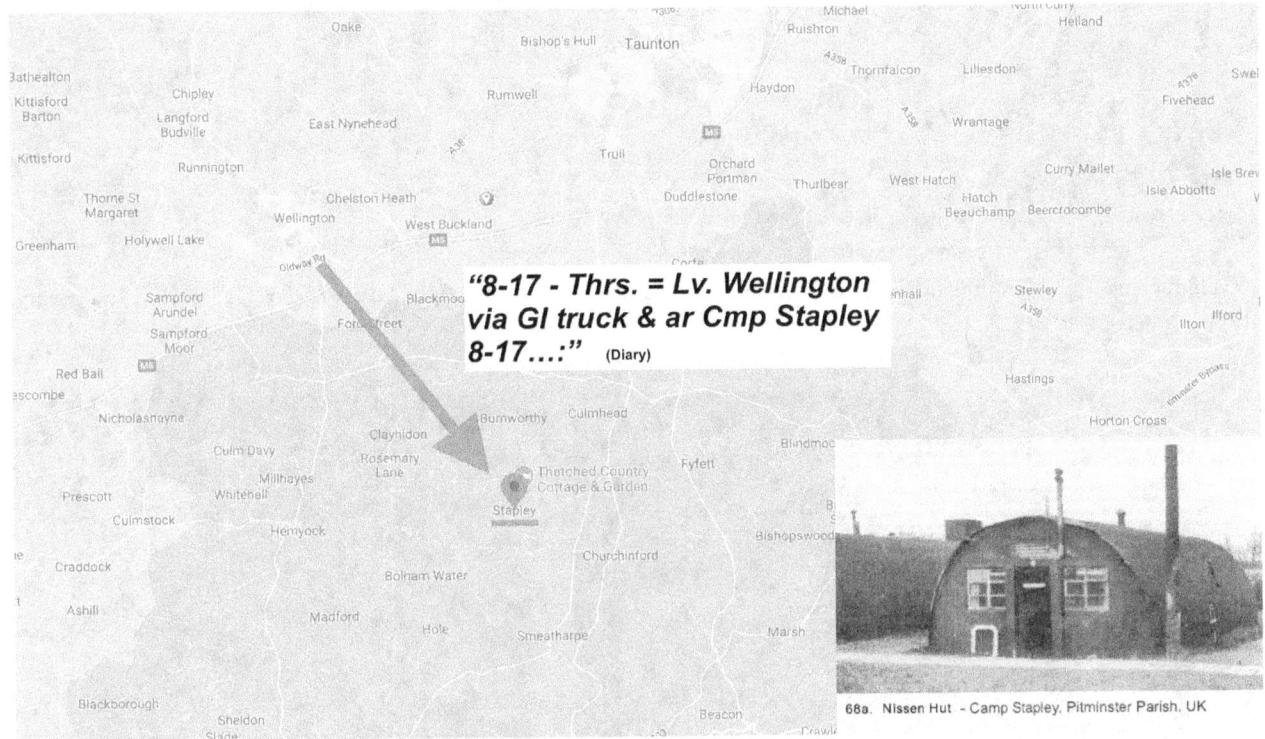

"8-17 - Thrs. = Lv. Wellington via GI truck & ar Cmp Stapley 8-17…:" (Diary)

66a. Nissen Hut - Camp Stapley, Pitminster Parish, UK

Left: Modern view of the fields north of the hamlet of Stapley where tens of thousands of American troop replacements encamped prior to being shipped across the Channel. Camp Stapley was one of three installations in the area, forming a kind of military triangle: Culmhead, an RAF base, was located less than a mile northeast of Stapley, and a few miles southeast near the village of Upottery, the US Army Air Force established a base from which the 101st Airborne launched its parachute drops on the towns of Normandy as part of the D-Day invasion.

New inductee reception center at Fort Meade that Ralph would have passed through. **Above Right**: GI's leaving (or arriving at) Fort Meade.

All that remains of Camp Shanks in Orangeburg, New York that saw 1.3 million GIs pass through its gates en route to Europe is the small museum seen above and the monument at right.

Above: The sprawling Camp Shanks complex in Orangeburg, NY. Doughs typically spent about a week at Shanks before shipping out. Ralph was there five days.

Troops arriving at Camp Shanks, New York. It is possible that Ralph is among them.

3. GI Orphan: Repple-Deppling Across France

Being a replacement is like being an orphan. You feel lost and lonesome. We want to feel
as if we are part of something, [but] as a replacement we are apart from everything."
— Infantry replacement[1]

On Sunday, August 20th, 1944, the call to leave came again and off we went by train to Southampton, a considerable journey. Unloaded at the outskirts of the city, we marched with full equipment through the city to the port and boarded the small English ship HMS *Devonshire.* We spent the night of Sunday, August 20, 1944, aboard the *Devonshire* in the Southampton harbor.

Early the following morning, this small ship made its rolling way across the choppy English Channel, arriving off the French coast of Normandy that afternoon. This was a boat ride never to be forgotten. I was sicker that one day than on the entire voyage across the Atlantic! Besides being seasick, I simply could not stomach the English chow in the kitchen (the coffee tasted remarkably like the dishwater which I believe it was), so I ate a D-ration (vitamin-packed chocolate bar) which made me even sicker.[2] That was the first and last of the much-touted D-ration for me. Good old K-rations were all right from then on.[3]

For some unknown reason (naturally it was unknown to us since we were never consulted about or advised on any move) we weren't allowed to disembark until around noon the following day, August 22nd. We went over the side of the boat, climbed down the side with all our equipment on a landing net and jumped for it into a landing craft — I believe the landing craft was a tank craft since it was merely an empty steel hull with a landing ramp at the front which was lowered when we hit the beach. We landed

---

1. Quoted in Klinek, p. 137.

2. The U.S. government had just four [requirements] for their new chocolate bars: They had to weigh 4 ounces, be high in energy, withstand high temperatures and 'taste a little better than a boiled potato.' The army didn't want the bar to be so tasty that soldiers would eat it in non-emergency situations. The final product was called the 'D ration' bar. As for taste, most who tried it said they would rather have eaten the boiled potato." (Stephanie Butler, "How Hershey's Chocolate Helped Power Allied Troops During WWII")

3. "A daily ration of three meals—breakfast, dinner, and supper—that gives each soldier approximately 9,000 calories with 100 grams of protein." (Richard Beranty, "K Rations Created the World's Best Fed Army," https://warfarehistorynetwork.com/article/k-rations-created-the-worlds-best-fed-army.)

on what was known as Omaha Beach[4] (each stretch of beach had been named prior to the invasion of June 1944, for identification purposes). Standing several hundred feet from the beach were rows of the craft that had been used in the invasion. Here at Omaha Beach, a very dear friend from Savannah, Orba John Bird, was stationed in the Navy and it is entirely possible that he was on the very same shore on which I beached, but at the time neither of us knew where the other was and so we didn't know to look around for each other.

The first several hundred feet of the beach was gradually sloping but then there was a sharp rise to about two hundred feet. Although it was now more than two months since the invasion, the beach was still a beehive of activity. Supplies and ammunition were stacked all over. The marks and scars of battle had not been obliterated and we were solemn as we slowly attacked the steep hill with our heavy equipment, thinking of the lives given in payment for this stretch of earth. Those of us who made the top at the first try got a welcomed break as we had to wait for the stragglers who, sick from the trip across and weakened from lack of sleep and proper food, could not make the grade.

After we got our wind, we marched about six miles with all our equipment until we were somewhere close (I was told) to the small town of Trevieres, France. This was the camp area of the 182nd Replacement Company, 42nd Battalion.

I had heard much about the lack of modesty in France, but I still could not suppress a gasp of amazement when, on my second morning in France, I awoke, crawled out of a hole, looked across the field and noted the outdoor toilet — a long wooden box with appropriate holes sitting over a slit trench and completely without camouflage. Five GIs perched on it like a limb full of buzzards while three French maidens calmly milked the cows within spitting distance!

No one ever bothered to tell us replacements where we were, where we were going or where we had been. I was not one to take this lying down, so every time a newspaper came within reach, I searched it for a map in an effort to keep up with our whereabouts. The first one I found I kept carefully folded among my writing materials and attempted as best I could to mark our route to the front. In addition to the maps, I kept a record of dates and places on scraps of paper.[5]

This part of France was as beautiful as England had been. It was largely a land of farms, cut up neatly by endless hedgerows. The very old church buildings were interesting, appearing much like the one on a postcard I picked up.

Our next move came on August 25, 1944,[6] when we were piled into trucks and moved to about eight

---

4. In the five months following D-Day, all replacement troops disembarked at either Omaha or Utah Beaches. This practice continued until November 10, 1944, after which Infantry Replacements arrived in Theater at La Havre, where the 15th Replacement Depot was located. These replacement depots were known in GI slang as "repple-depples." (Lerwill, p. 453).

5. See Appendix to view Ralph's hand-written, day-by-day chronicle of his movements, locations and experiences from the time of his mustering in in September of 1943 in, to the day of his departure for the States from Le Havre, France on July 12, 1945.

6. While Ralph was being trucked to Le Mans on August 25, his future unit, the 4th Infantry Division, was marching triumphantly down

miles out of Le Mans, France (485th Co., 87th Bn., 14th Regt., another Infantry Replacement outfit). I guessed this trip to be about a hundred miles. The weather was beautiful and warm. Apple and pear trees were loaded with fruit but not yet ripe. Sleeping on the ground had so toughened me that even chigger bites (I once counted nineteen behind one knee) were shrugged off with little concern. Bob Merz occupied my pup tent with me at this place. It was the last day of our stay that I bit off more than I could chew one day at lunch. Bees swarmed everywhere, especially at lunchtime around the gallon cans of orange marmalade. One got on the underside of a piece of bread and into my mouth. After it bit me, my mouth became so swollen I could open only a small part of it.

On September 1, 1944, it started to rain and turned really cold. Naturally, this was the signal to move. We left about noon via truck and arrived at Mortagne, France (an estimated fifty miles northeast) at dusk — 234th Co., 90th Bn., 3rd Repl. Depot, APO 153.

This encampment near Mortagne was on a high embankment. It was here that I ate my first hard French bread — huge improvement over the "bread" tossed onto our plates at the mess tent!

THE RAINS CONTINUED WITHOUT LETUP. The fields became a solid mass of mud. MUD — wet, clinging, cold, nasty, ever-present MUD. We hated it more than we hated the Germans. An article I clipped from *Stars and Stripes* ("Mars Name Is Mud") perfectly described the horrible weather — and the even nastier conditions it produced — during my time in France. The misery of being an Infantry Replacement was made even more miserable by the endless rain and seemingly bottomless mud we had to camp in. Mud — the greatest enemy of an advancing army.

At 1:15 am Friday, September 8, 1944, the trucks came to haul us another one hundred miles to Melun, about thirty-three miles southeast of Paris. We arrived at 9:00 am and left Melun at fifteen minutes after midnight the next morning, September 9. The front-line guys were pushing the Krauts hard and fast back to the Fatherland.

Before we left Melun, however, there came an event that both surprised and delighted all of us. In a huge field nearby, Bing Crosby and his troupe held a show for us. I luckily got a "down-front" seat (my helmet) and could hear most of it. It was quite a blow to me that Bing was nearly bald — a fact that had heretofore been a carefully guarded secret — at least from me! It was the same old debonair Bing who talked about "Hairless Crosby" and "Hopeless Hope" and "Breathless Sinatra."

During the course of the show, there passed overhead a B-29 bomber with part of one wing shot away and one engine. It was a scary reminder that we weren't in Hollywood, but in the middle of a very real and very deadly war.

This show went a long way to make life more livable and was encouraging to say the least. It was the only "name" show I saw overseas, although I heard there several others.

---

the Champs Elysees in Paris, having liberated the city that same day.

I vividly recall that departure from Melun at 12:15 am, September 9, 1944. It was so dark that we weren't sure everyone was on the trucks or not. Some of the boys were pretty heavy sleepers and we didn't have but a few minutes to roll up our stuff. As far as I know, though, no one was left behind. While en route, our convoy halted for a time — we learned later from a local that four Jerries had stopped the trucks. We must have captured or killed the Germans because we soon were on our way once more.

We neared Soissons at 9:00 am, September 9, 1944, by-passing Paris, and camped there. I'd guess it was a ride of about eighty miles. The dull thuds of bombing and artillery were heard until the front moved on, across the Belgian border some seventy miles east of Soissons.

"My, sir—what an enthusiastic welcome!"

While at Soissons, I became close friends with Joe "Ralph" Behan, Bill Maddox from Indiana, Don Adams, "Boots" Bizianes, a lanky Kentuckian, and Walter J. Blazinski, a New Jersey Polack. Some of us took a seven-or-so-mile hike while we were encamped in Soissons; we saw how badly the Barisis area north of Soissons had been bombed — the rail line was even more severely damaged by the raining allied bombs. Before we left Soissons, Adams and Behan got shipped out — lucky devils!

I learned from the men of the 4th Infantry Division which I was later to join that they had received a joyous welcome in Paris, while those of us who were mere replacements awaiting assignment were hauled miles around Paris to the south — reminded me of Mauldin's cartoon above.

Above: I cut out this *Sad Sack* cartoon after I had my first experience "taking care of business" in France — "sightseers" indeed! Of course, I would soon come to realize that this kind of thing, as unnerving as it was to a newly arrived IR, would be the least of my worries in places like the Hurtgen Forest and the Ardennes.

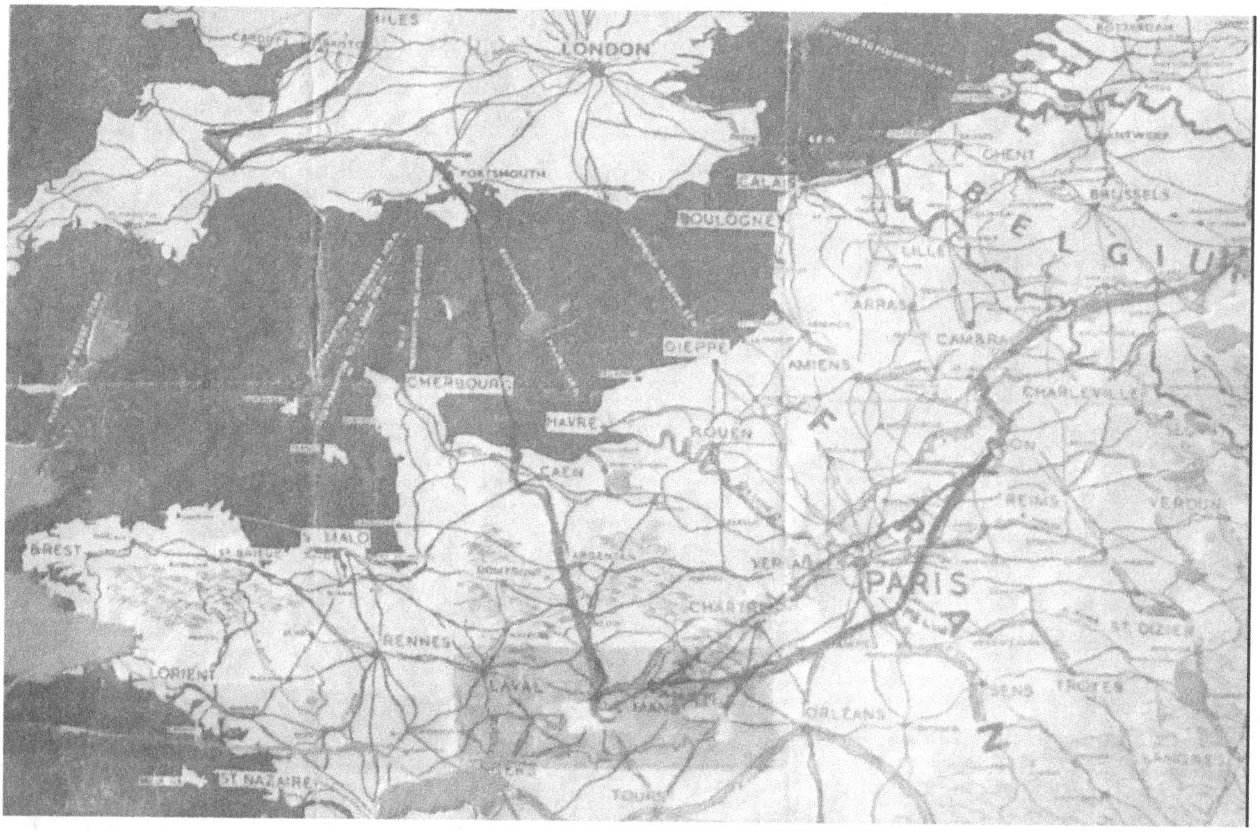

'Mars' Name Is Mud

THESE men are making their way from bivouac to chow line. Because they are at war, they naturally plod through mud. We don't know what goes wrong with nature's plumbing when the guns begin to shoot, but something always does.

Mud dragged at the frost-bitten feet of Washington's Continentals. The red mud of Virginia sucked at the cannon wheels of the Army of the Potomac and at the hooves of Jeb Stuart's cavalry. Soupy mud engulfed the trenches of 1917-18.

We say the traditional presentation of Mars, god of war, has never been accurate. He ought always be shown with his feet and hairy shanks buried in mud. In fact we'll go further than that. We say that whenever history books discuss war, mud should be mentioned. And no history book should show a war picture that doesn't include mud.

Only when everybody on earth is made to realize that mud instead of trumpets and banners is the true symbol of war is the world likely to have perpetual peace—or even a reasonable

One big trouble with the Germans as a race is that nobody ever tells them about the mud. They only hear about the trumpets and banners. One way or another, it's up to all of us to show Germany that war means mud and things like it.

It's a fact that must be kept before the German eyes for as many genera-

trumpet and banner pipe dream out of the race. The rest of the world has got to arrange its affairs so that when any future Hitler rises up to blow a trumpet and wave a banner he can be—figuratively speaking—hit square in the kisser with a juicy mud pie.

P.S. Beat the mud! If you have any ideas on how to be mobile in spite of mud, send them to the Beach, Stars

Clockwise from top: (1) The map I clipped from a newspaper shortly after landing at Omaha Beach on which I attempted to track my movements as an IR. (2) Postcard I picked up depicting the kind of churches we saw across France and Belgium. (3) Mud — bottomless quagmires of the stuff plagued us everywhere we went, or so it seemed. It made our lives about as miserable as the Krauts did.

Left: *Stars and Stripes* article I kept about Bing Crosby's shows for US troops in France. I saw his show while in Melun, France.

Left: Rain and mud continued to plague us throughout our movements across France. This picture of an exhausted GI that I clipped from *Yank* magazine pretty well captured how we all felt about it.

* * * * *

Left: The HMS *Devonshire,* the small British ship that wallowed and rocked its way from Southampton to Normandy, where it deposited its human cargo of Infantry Replacements onto Omaha Beach August 22. The IR's, many green with seasickness from the rough journey across the Channel, staggered through the scattered detritus of D-Day and up the high, sharply sloping bluff en route to an IR depot near Trevieres. Their path would take them through ground that today is guarded by seemingly endless ranks of silent sentries — the gleaming white headstones of thousands of slain American GI's at the American Military Cemetery near Colleville (**see map below**).

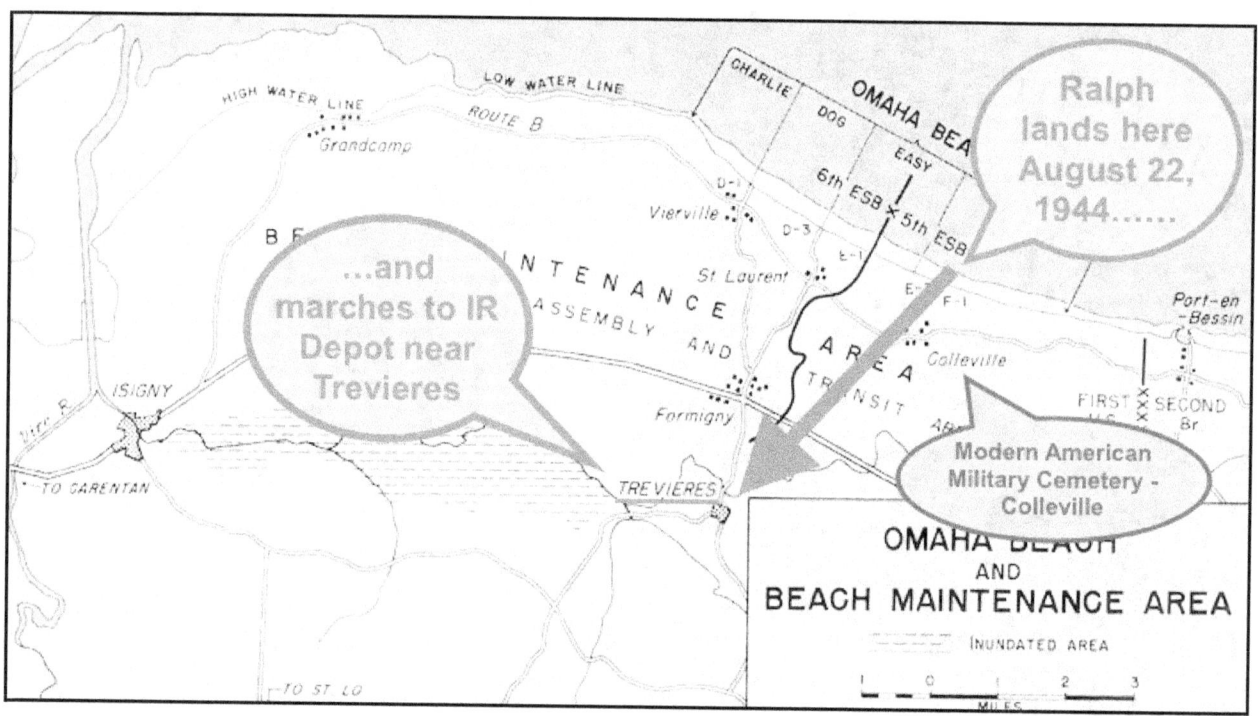

(Map source: https://en.wikipedia.org/wiki/American_logistics_in_the_Normandy_campaign)

Omaha Beach, Normandy,
France

Omaha Beach, Normandy, France
Likely spot where Ralph disembarked from HMS
Devonshire August 22, 1944

Left: The 200-foot bluff rising from the sands of Omaha Beach scaled by Ralph and his fellow exhausted, sea-and bad-food-sickened GIs on August 22, 1944. The climb was physically demanding and emotionally sobering as the IR's contemplated the bloody price that had been paid ten weeks earlier for the ground they traversed.

Left: Modern view of **Trevieres,** site of Ralph's first IR (Infantry Replacement) Depot encampment — 182nd Company, 42nd Battalion. The IR's remained near Trevieres from August 22 until August 25. Following behind the advancing lines, the IR's ended up at Le Mans, 130 miles southwest of Paris.

August 22, 1944: "We marched about 6 miles with all our equipment until we were close...to the small town of Trevieres, France, the camp area of the 182 Replacement Co., 42nd Bn."

Above: Broad fields eight miles west of **LeMans,** France, where Ralph and his fellow replacements arrived August 25, 1944, the same day that the 4th ID liberated Paris. Several replacement depots were located near Le Mans, "where excellent highway, billeting, and training facilities were available." The Ground Force Replacement Command (GFRC) Headquarters were moved to these facilities on August 23, 1944, just two days before Ralph arrived (Lerwill, p., 447). Ralph's IR unit was the 485th Co., 87th Bn., 14th Regiment.

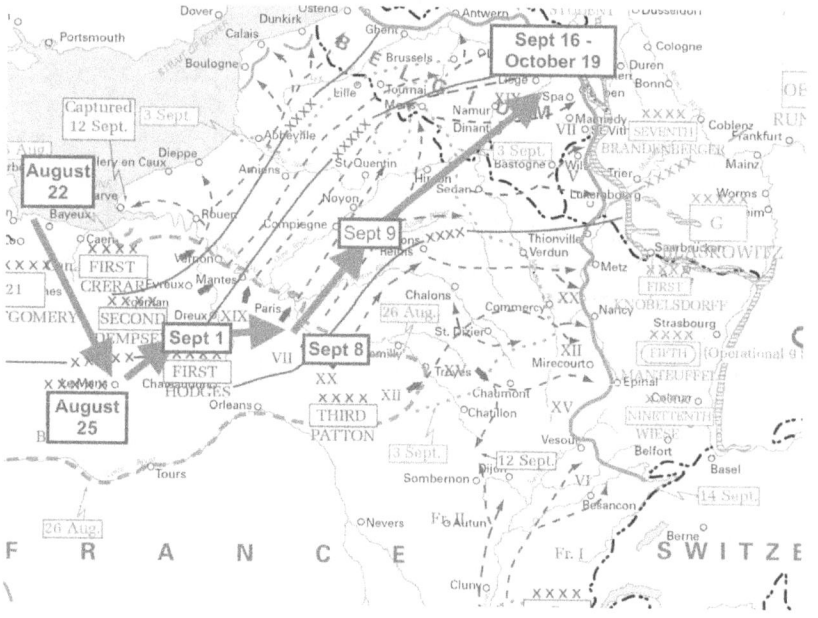

Left: The more prominent arrows trace Ralph's movements as an infantry replacement in relationship to the advancing Front. At each stop along the way, a number of "repples" would get their "call" to go to the front lines to replace ever-mounting casualties. The Front had stabilized along the Belgian-German border as Ralph arrived in the vicinity of Verviers, location of a major IR Depot southeast of Liege, Belgium — it was here that Ralph got his "call" to the 12th Regiment of the 4th Division on October 17, 1944.

(Map source: https://www.westpoint.edu/academics/academic-departments/history/world-war-two-europe)

Captured airfield, Melun, France
1944 —
Possible Infantry Replacement
Depot location

"In a huge field, Bing Crosby and his troupe put on a show for us...during the show a B-29 bomber flew overhead with one wing shot off and one engine dead."

Left: Ralph was at this location for less than twenty-four hours, from 9 am September 9 until just after midnight September 10. But it was a memorable few hours as he got to see Bob "The Crooner" Crosby in person!

Below: Photo of area near Mortagne that matches Ralph's description of the location of the 234th Co., 90th Bn., 3rd Repl. Depot on September 1 – 8, 1944.

1 - 8 September 1944: "We were encamped on a high embankment near Mortagne"

Open fields two miles north of Soissons
General area of Infantry Replacement
Bivouac September 1-8, 1944

Left: Probable vicinity of IR encampment north of Soissons. I calculated this location from Ralph's comment about having hiked about seven miles north to Barisis from his camp near Soissons.

Above: *"Some of us took a seven-or-so-mile hike while we were encamped in Soissons; we saw how badly the Barisis area north of Soissons had been bombed — the rail line was even more severely damaged by the raining allied bombs."* This photo of the Barisis rail station was taken some years before the War.

4. Joining the Furious Fourth

"... the new recruit is treated with as little civility as possible, but it was so good to be at long last assigned to a unit, little else mattered."

In a very long convoy we left Soissons, France on a Friday night, September 15, 1944, at 9:00 pm. Through the rain and long night, we journeyed about 125 miles across the rest of France and most of Belgium to a point about twenty miles west of Liege, Belgium.[1] Forty miles further east lay Aachen, Germany. The American advance beyond the Siegfried Line had been blunted and the front became stable.[2] We sat here near Liege for a whole month, just waiting. Our first night at this location, a plane dropped a flare about a hundred yards from my hole and nearly scared me to death. We heard that a convoy had been badly strafed twenty miles away and we all had the jitters.

One night, in spite of the rain, we were to have a movie — *Going My Way*, starring Bing Crosby. We saw some of it, but it was interrupted in the middle by enemy planes. This was to be only the first of a series of incidents which kept me from seeing this film all the way through!

To our little close group — Miles, Maddox, and Blazinski — we added Howard Calkins of Cleveland, Ohio. He was not only a quiet, sensitive, intelligent, and educated fellow, but he also had a camera, so we all vied for the opportunity to get our photos made during our month-long "hurry-up-and-wait" outside Liege. During this time, Calkins got his "call," and I later heard he landed a pretty good job in Paris. I acquired another friend, John Nivala of Detroit, Michigan, who had taken basic training with me at Fort McClellan.

I recall several things about our time at Liege rather vividly. Throughout our stay, the noise of battle

---

1. Ralph and the other IR's were, more precisely, southwest rather than west of Liege, based on Ralph's diary entry recounting his departure from this area a month later: "10-17 Tue — Lv Abee Noon." Abee is a subdistrict of the village of Tinlot, located about 20 miles south-southwest of Liege.

2. There was, in fact, a good deal of action taking place to the east along the West Wall. The 4th ID had pushed across the West Wall on September 12, and engaged in fierce fighting in the Schnee-Eiffel region; their efforts were blunted by strong German countermoves and the division withdrew back into Belgium. The 4th would re-visit this bloody ground the following winter with greater success. In addition, beginning October 2, other US forces were attacking through the Aachen corridor, finally taking that city in bloody urban fighting on October 22.

came to us closely. Daily, vast numbers of bombers roared overhead toward Germany and returning — at least most of them. It was clear that we were getting quite near to the front; one Saturday, I did my KP duty to the steady thunder of artillery and bombs. I nearly stabbed myself with my paring knife when one blast landed scarcely a thousand yards away from me!

September 23, 1944, was a red-letter day. Early in the morning we were pushed out of the woods into the foot-deep bog that used to be a field. It was cold, ugly, and miserable. We were told to lay out all our equipment; some GI's had gone AWOL to the city of Liege and the brass wanted to see who was missing and who might have stolen the missing men's equipment. Then it began to rain — hard. We stood all day long in the driving rain for no earthly purpose that I have yet been able to discover. About thirty minutes before dark and after every piece of equipment was thoroughly drenched, we were ordered back into the woods. Yeah, rough.

In October, Germany's V-1 rockets, better known as "buzz bombs," rumbled westward overhead toward England. These noisy flying bombs had wings and a jet exhaust protruding from the dorsal side. Sometimes we would see the engine quit and the bomb would take a nosedive toward the earth, exploding near us.

Monday, September 25, 1944, was a great day — I received my first mail from home since England! Four whole letters! To the itinerant unassigned replacement, there were three major problems — in order of importance:

1. *Mail from home.* As we were moving so constantly, there was little chance for mail to catch up. As I read back over the letters I wrote home, the most constant theme was the lack of mail.

2. *When will I be assigned?* Those who have never wasted away in a replacement camp will never be able to understand the depth of such a question. Your friends are called away one by one, and you make new ones, only to lose them too. I've heard about people who "found a home in the Army," but surely none of them were ever replacements.[3]

3. *What's going on?* We spent much time going from place to place but seldom knowing one from the other. The occasional paper that fell into our hungry midst was worn as soft as toilet paper (for which it was always ultimately used).

Our next ride began October 17, 1944, at about noon, lasted for three and half hours and ended at

3. "Being a replacement is just like being an orphan. You are away from anybody you know and feel lost and lonesome. We want to feel that we are a part of something. As a replacement we are apart from everything. You feel as if you were being pushed out of a place blindfolded. You feel totally useless and unimportant. They treat us like idiots, and we don't disappoint them" (GI in an IR camp, September 1944, quoted in Klinek, p. 137).

3:30 pm in the Ardennes Forest near Malmedy, Belgium: 310th Co., 41st Bn., 3rd Infantry Replacement Depot.[4] We were about thirty-five miles south of Aachen, Germany, and fifteen miles inside Belgium. This was a thickly wooded area that seldom if ever saw the light of day. The ground was softly padded with many years' deposits of needles. We spent one night here. The next day, it would be my turn to leave the "boys."

October 18, 1944, a heavy Colonel with a cane took a sizeable group of us out into the open (and into the rain) with all our equipment and proceeded to deliver a dissertation of all the "firsts" of the Famous Fourth Infantry Division. We were informed that this outfit, positioned nearby, would be our future "home" and we would be part of the big push into Germany proper. The build-up was largely lost on us because we'd already been through so much it didn't matter a tootle which outfit we joined or what it had done or would do — we were ready for a change. We were marched off, piled into trucks, and taken to the nearby Service Company, 4th Infantry Division, 12th Infantry Regiment. John Nivala was left behind and I never heard from him again.[5]

It was pitch dark when we arrived at Service Company and raining like mad. A GI I couldn't even see, and I buttoned our shelter-halves together by feel, "felt" out a place on the ground to pitch it and turned in, all in utter blackness and in the rain. My gear lay outside in the rain soaking it up, but I didn't care then. I was, at long last, about to get located.

Holzheim, Belgium, October 19, 1944, Headquarters and Headquarters Company, 12th Regiment, Fourth Infantry Division:

It was a sad looking replacement who staggered into the upstairs room (first house I'd been billeted in since the States) occupied by part of the Radio Section, 12th Infantry Regiment Headquarters.

As I stood in the doorway, dripping from every crease, the guys took one look and recoiled in horror. I can see them now, sitting around a table that jutted out from one end of the room, playing cards. Most of them were Pennsylvanians[6] with only one other Southerner in the group — a 6 ½ foot North Carolinian by the name of Ernest Jarrett. Most of them had been together since the unit had been organized in

---

4. The enormous 3rd Replacement Depot was located fifteen miles northwest of Malmedy, near Verviers, Belgium. The depot provided replacement troops and direct support for the US 1st Army.

5. Klinek recounts a similar experience described to him by an IR assigned at about the same time and place as Ralph: "... We learned we were joining the 26th Regiment [1st Infantry Division]. A captain told us briefly about the formidable record of the Division and the Regiment ... His talk was simple ... Names were then called out. The men were assigned to different battalions, thence to companies, platoons and signals" (Klinek, p. 275).

6. Charles Emery (Erie), Raymond Kimmel (Lebanon), Robert Nace (Minersville), Nunzio Yocca (Windber) and Emil Lukert (Philadelphia) — all told, more than a quarter of the Radio Section hailed from the Keystone State.

the States. As is quite customary I found out, the new recruit was treated with as little civility as possible. I keenly felt their aloofness but went about quietly to get myself settled. It was so good to be at long last assigned that little else mattered.[7]

The next day I took a look through our company area and ran across a guy I took radio training with at Fort Benning — two of them in fact — Tom Norton and Stanford Blose. Blose (everyone in the Army has only a last name) was an excellent radio operator and technician in civilian life, so you want a guess what he was doing? Pulling guard duty. But me, an inexperienced novice — a radio operator in Regimental Headquarters! That, my friends, is how it goes in the good old U.S. Army.

As the buzz bombs flew continuously overhead, a small Italian in our group named Verga built a small receiver from the remains of an SCR-536 (Handy-Talkie) and we were able to listen to German propaganda stations and BBC (that is, when we could get Verga to quit fooling around with it trying to make it better all the time and only succeeding in making it worse). A few days later came all my backed-up mail — Happy day!

Unbeknownst to us at the time, we would soon encounter much worse grief and destruction in a horrible forest called the Hurtgen.

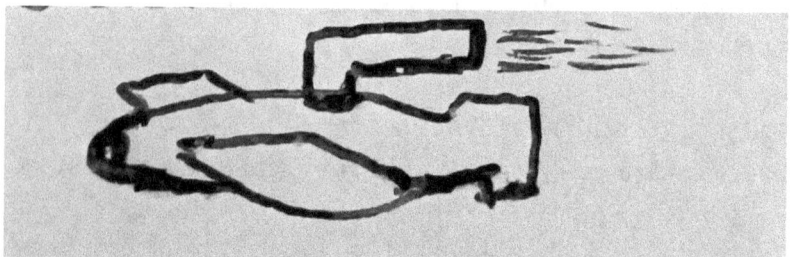

Above: A sketch I made of the German "buzz bombs" that flew overhead constantly during my time in the 3rd IR camp outside Leige.

---

7. "Though the replacement finally had a home, not all of his new comrades welcomed him with open arms. Many replacements complained that they received a frigid reception from veterans…. and replacements frequently lamented that it took quite a while until they truly felt that they belonged to that team." (Klinek, p, 136). In Ralph's case, that feeling of full acceptance did not happen until after he had proved his mettle in a devilish place called the Hurtgen Forrest.

Led by Love of Country | 47

Howard Calkins (**left**) and me near Liege, Belgium in September of 1944. Calkin got his assignment call while here– a cushy Paris assignment. I would not be so lucky.

Mail from home was pure gold, especially for the always-on-the-move infantry replacement. During my time as an IR, I felt a lot like Private Breger (**left**).

Above: Some now-worthless coins and paper money I picked up while advancing through Belgium, Luxembourg, and Germany. The coins are genuine, as is one of the bills. The rest are "invasion notes" the American government issued to us in each country we occupied.

Mauldin's cartoon (**above**) reminded me of those days of being without a "home" in the replacement camps.

Below: These insignias adorned my uniform from October 19, 1944, until October 14, 1945 — one year almost to the day.

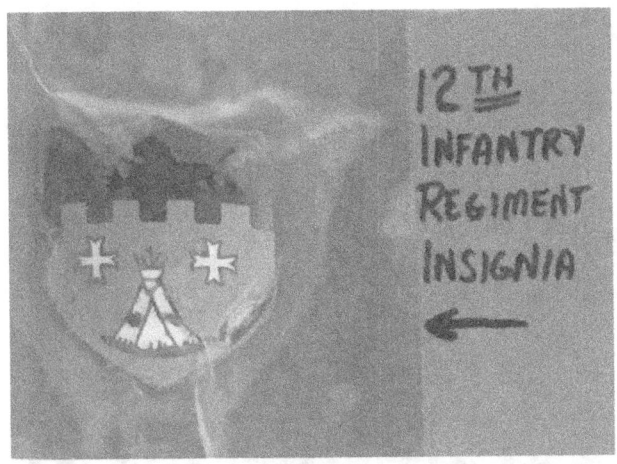

Mauldin's cartoon (**above**) accurately captures those occasions when a German or Belgian civilian would come around and look reproachfully at us for taking over their houses and for the severe damage done to their homes and villages.

Abee
District,
Tinlot,
Belgium

Above: Area where Ralph encamped as part of a temporary Infantry Replacement depot from September 16 until October 17, 1944. **Below:** Map showing the location of Abee, a district of the village Tinlot, where Ralph remained from September 16 until October 17, 1944

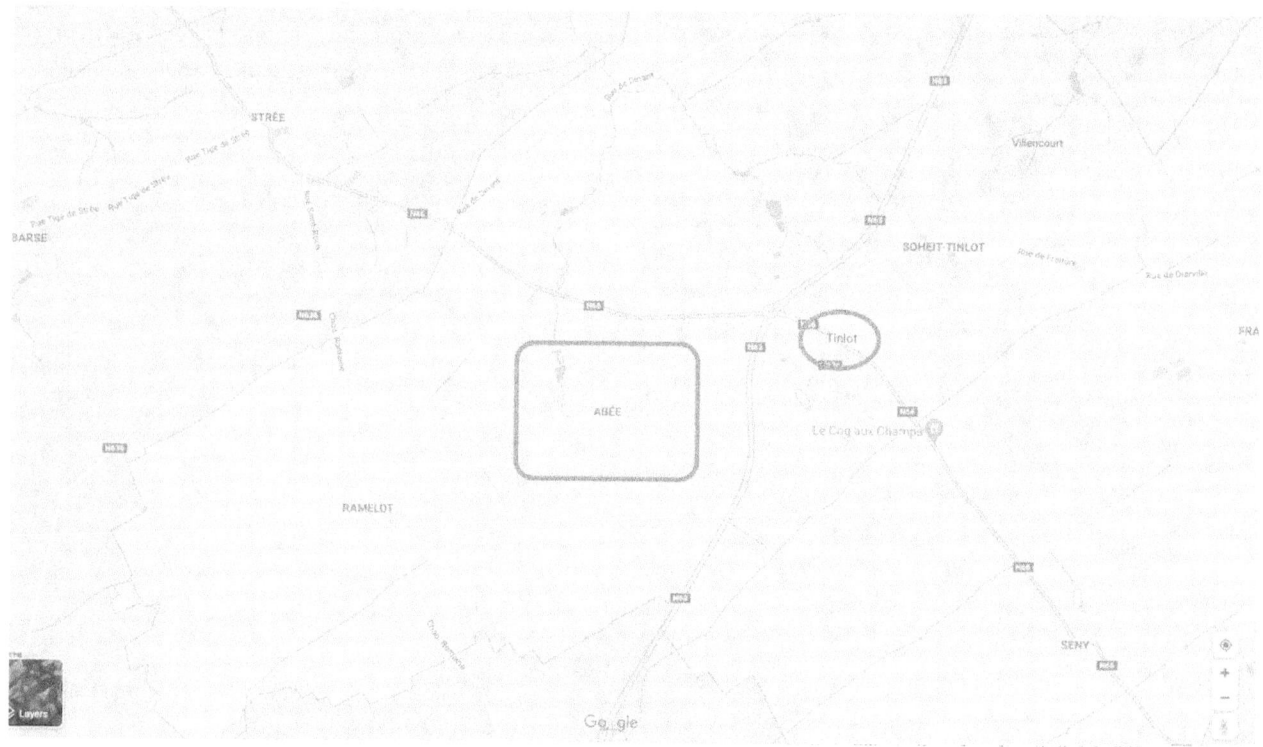

5. Into the Meat Grinder: The Hurtgen Forest

"Hurtgen was agony, and there was no glory except the glory of courageous men"
— **Mack Morris, Infantryman, U.S. Army**

All good things must come to an end, and our long stay at Holzheim ended on November 6, 1944. We set out at six in the evening in a cramped half-track for an all-night trek. Traveling without lights was uncomfortable, to say the least, especially when the moon came out to reveal we were winding around mountains with a sheer drop-off on one side most of the time! The convoy lost several tanks, trucks, jeeps, and other vehicles because they tried to hug the side of the mountains too closely and mired up. Finally, fitful sleep overtook most of us until it started to sleet. We jumped out, pulled the canvas covering over us and promptly went back to sleep.[1]

When I awoke the next morning, my legs and back were about to kill me. The canvas top had thrown us all together in a pile in the half-track and it took some doing to extricate myself. When I crawled out (we had stopped finally around midnight) and saw where we were, there was a great temptation to crawl right back in again. Woods — everywhere woods! Nothing but trees. Heavy, thick, foreboding woods. I later learned we were in the Hurtgen Forest, just south of Aachen, in Germany. The sun couldn't shine through the thick forest overhead. The trees looked like a cross between pine and fir. They had been planted and remained in almost perfect lines. There was very little underbrush — just enough to hide a million mines. And this forest was to be "home" for the next month!

* * * * *

Unknown to Ralph and his cold, rain-soaked comrades, the 12th Infantry Regiment had been detached from the 4th Infantry Division and temporarily attached to the 28th Infantry Division whose battered 109th Regiment the Ivymen relieved, unit by unit, foxhole by foxhole, in the dead of a bitterly cold and

---

1. Ralph's account of this harrowing forty-five-mile journey is confirmed in the official 12th IR after-action report for November, 1944, dated 4 January 1945: "The entire march was made during the hours of darkness. The night was intensely dark and very rainy... muddy, narrow roads caused vehicles to slide into ditches... the blackness of night hampered operations." Also, official accounts have the regiment moving out at 1410 hours — 2:10 pm. Regimental CP personnel — including Ralph — did not depart until four hours later.

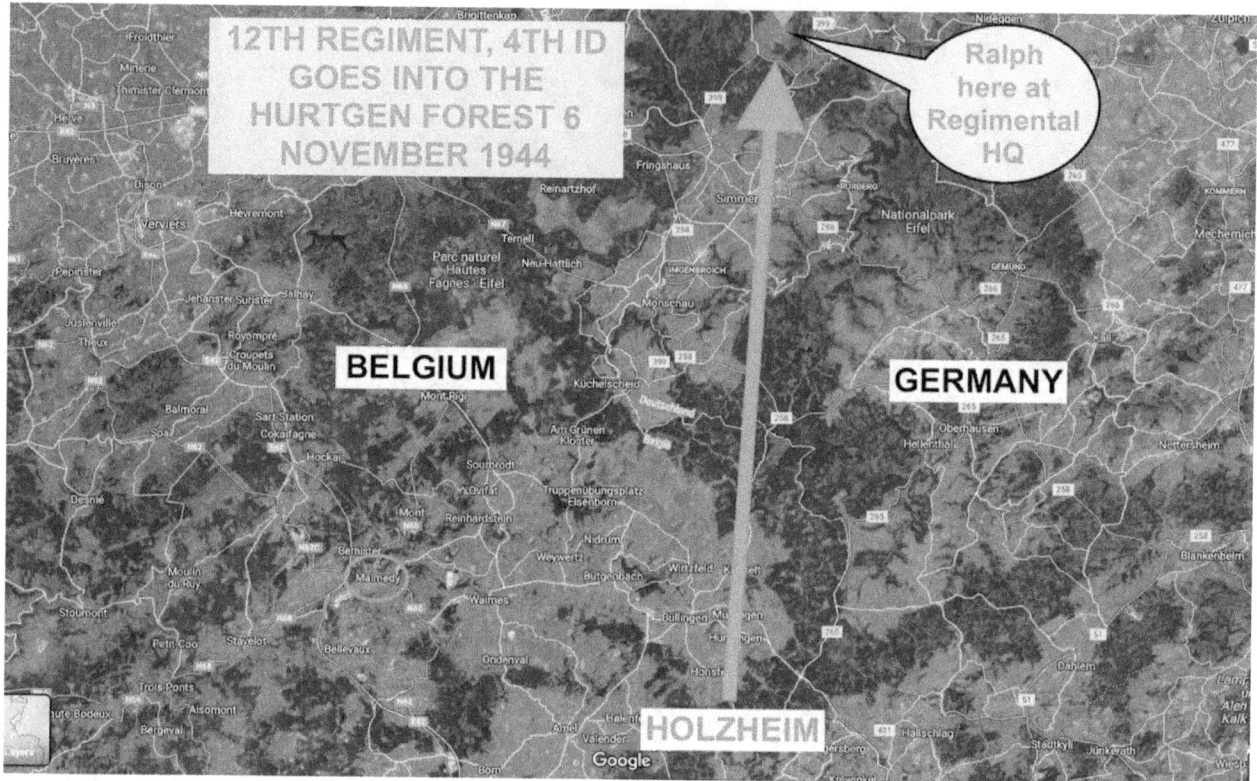

stormy night without any prior reconnaissance — a truly remarkable accomplishment in and of itself. The 28th had been so badly mauled during its time in the Forest, that the original plan to send the Fourth into a reserve position further north was scrapped and the 12th — already enroute to its original assigned destination — changed course in the dead of night under radio silence and turned eastward toward Germeter to relieve the shredded companies of the 109th. Its two sister regiments — the 8th and 22nd — arrived a few days later. By the evening of November 10, the Fourth was whole once again — but not until the 12th had already undergone a horrific baptism of fire during the deadliest five days of fighting in regimental history, from November 7 to 11.[2]

This foolish — and inexplicable — practice of introducing units piecemeal into the Forest typified the tactics of the brass from the very beginning of the battle, and the grunts on the front lines paid dearly for their commanders' tactical myopia.

The stated objective was to clear the forest and beyond in order to capture the dams on the Roer River, which the Allied Brass feared the Germans planned to dynamite in order to flood the Cologne plain

---

2. "Condolence Letters - Lieutenant John W. Irvine, 4th Infantry Division (1944)," retrieved from http://omnigatherum.ca/wp/?cat=35; q.v. Johnson, pp. 196-199. q.v. Memo: Headquarters 12th Infantry APO 4, US Army, Subject: Action Against Enemy, Reports After/ lAfter Action Reports. To: The Adjutant General, Washington, 25 D.C. 4 January 1945. History of the 12th Infantry Regiment for the Month of November 1944.

and impede the American advance. A better plan would have been to advance through a relatively open corridor fifteen miles south of Aachen, but military planners feared a German ambush on their right flank.[3] Ironically, it turned out that the Germans had no plans to blow the dams at all.[4]

The only true positive to the whole operation was the incomparable endurance and heroism of the Yankee GI's, who clawed their way yard by agonizing yard to the eastern edge of the Hurtgen at a cost of 35,000 lives over three horrifying months of blood and gore. A Pyrrhic Victory if ever there was one, the entire Hurtgen campaign is viewed by military strategists and historians as a needless and colossally tragic tactical blunder.[5, 6]

As regimental historian Gerdon Johnson eloquently observed, *"Hurtgen Forest is a name to carve some day on the war memorials of America beneath such evocative place names as Chateau-Thierry, the Argonne Forest, and the Wilderness in the War Between the States. Other battles have been more dramatically decisive… but none was tougher or bloodier than the battle for this Hurtgen Forest."*[7]

Sadly, despite its dubious distinction as the bloodiest single campaign of the war and the longest battle in American military history, the Herculean effort of the American soldier in the Battle of the Hurtgen Forest has been largely overlooked in popular accounts of the war — due largely to its being overshadowed by the more strategically decisive Battle of the Bulge that broke out on December 16, 1944. In our travels in the region, we found that even the Germans who live in the area today have never heard of the Battle of the Hurtgen Forest.

Of course, like all the other dogfaces, Ralph knew none of this. He — and they — went where they were told to go, did what they were told to do, shot at whomever was shooting at them, and tried to live to fight another day…

At this point, we re-join Ralph's account…

* * * * *

Robert T. Weltzein (my now best Army buddy of Seattle, Washington) and I set about digging a hole

---

3. https://en.wikipedia.org/wiki/Battle_of_Hürtgen_Forest. q.v. Rick Atkinson, "The Hurtgen Forest: The Worst Place of Any" on https://www.historynet.com/the-hurtgen-forest-1944-the-worst-place-of-any/.

4. https://en.wikipedia.org/wiki/Battle_of_Hürtgen_Forest.

5. "Condolence Letters - Lieutenant John W. Irvine, 4th Infantry Division (1944)," retrieved from http://omnigatherum.ca/wp/?cat=35. See also C. Peter Chen, "The Battle of the Hurtgen Forest 19 Sep 1944 — 1`0 Feb 1945," at https://ww2db.com/battle_spec.php?battle_id=117.

6. "…a misconceived and basically fruitless battle that should have been avoided," wrote Charles B. MacDonald, US Army historian and Hurtgen survivor (quoted in https://en.wikipedia.org/wiki/Battle_of_Hürtgen_Forest).

7. Johnson, p, 195.

about three feet deep, four feet wide and six feet long. Then we cut down trees, cut them into lengths, and laid them over our hole in a double layer. This was to protect us against tree bursts, our worst enemy outside of trench-foot and pneumonia.

Weltzein and I finished our hole about dark. Along came Shorty Walters begging for a safe place to sleep! He's a good Joe so in he comes. We piled the bottom of the hole with a layer of branches, padded the dripping cold clay sidewalls with our raincoats, put two blankets on bottom and all the rest on top. I don't suppose the temperature was more than five degrees above zero. We were well clothed for the job — starting from the bottom: cotton underwear, woolen underwear, wool OD (olive drab) uniform, fatigue uniform, wool knit sweater, field jacket, overcoat, wool knit cap, helmet, two pairs socks, combat boots and overshoes. To prepare for bed, we stripped off the overcoat, field jacket, and overshoes. The next morning there would be a little matter of six inches of icy water in the hole, in our helmets, and in the blankets, drenching us to the skin. Every day it was to bail out.

Being the newest guy in the outfit, the first call for a volunteer (that's an Army word, not at all like Webster's) to go to the front for substitute duty had my name on it. I had to untangle my blankets and things from those of Weltzein and Walters, board a jeep and take off in the night. The jeep driver stopped on the side of a mountain and said I'd find the company around the side of the mountain but that he wasn't going any further. There was nothing for me to do but stumble on around there, which I did, frozen more with fear than the cold. I found half the company hovering inside a captured Heinie pillbox taking turns getting warm, the rest holding the fort outside. There was no room for me inside, so my SCR-300 radio kept me company outside!

The radio and I hugged the side of the cement pillbox as German shells exploded every few minutes on the opposite slope of the mountain. As the night wore on, it got colder and colder and I continuously beat my hands and feet to keep them from freezing. That night was a million years long. When I was still alive the next day, I felt like Mauldin's Willie — "a fugitive from the law of averages"! (**see page 61**)

Weltzein, Walters and I became permanent foxhole partners during this battle in the forest. While we took turns staying outside to see if any light came through, we managed several workable lighting and heating systems! The first was a small food tin filled with stolen gasoline, some rags, and, presto, light! The stench of it filled our eyes, the smoke choked us, and the shelter-halves we had stretched over the logs to help stop the rain became so blackened with smoke you could still rub some off when I turned it in back in the USA several months later! Later, as packages came from home, we got candles. By the light of these we were able to write home most of the nights when we weren't on the move. We wrote lying flat on our backs with the paper pressed against the underside of the logs which constituted the top of our lair.

I vividly recall the time when we purloined a large can of orange marmalade from the mess outfit and ate ourselves sick. I seldom see any in the stores today but that I don't think of it and vow never to eat any more. We became very close in those hours in the night, exploring all the usual range of subjects — women, home, religion, war, fear, food.

Another event I vividly recall during the Hurtgen Forest fight was the time when I was told I had volunteered to operate a radio in an outpost across the river[8] from the Germans. There were four of us in this detail — three wire section linemen and me. We dug a big hole, but before completion we were being shelled. All four of us tried to get in the hole, but I couldn't squeeze in. I sought shelter under a fallen tree.

With my portable radio on top of the tree, I could make contact with the rear, but heavy mortar fire kept me pinned down under the tree. I dug out a shallow hole while a million spiteful bugs explored my anatomy from there to you-know-where. But it was better inside than out, as I discovered the next morning when I reached out into my case of rations, only to find it riddled with shell fragments. The location became too hot so were evacuated after two days. Sergeant Woodrow Bledsoe of Altoona, Georgia, was the first to try to rescue us, but he got a direct mortar hit on his jeep. I yet do not know whether or not he made it.[9]

Thursday, November 16 was a beautiful, cold clear day — perfect for us to launch an attack at 11:00 am.[10] The Krauts had done a devilishly good job making the forest all but impenetrable to our advance. Minefields, entrenchments, pillbox fortifications, not to mention the natural impediments of a dense forest. Casualties were extremely heavy in the Hurtgen. So much so, in fact, that it was extremely difficult to get on "sick call."[11] Especially fearful and demoralizing were tree bursts. Powerful German 88's hurled shells into the tops of the tall firs and pines where they exploded, showering all of us down below with

---

8. I.e., the Weiser Wehe River, which runs from north to south through the 12th Infantry Regiment positions. It lay due east of and directly adjacent to Raffelsbrand where Ralph was with HQ & HQ company, November 7-21. See satellite map and modern-day pictures at the end of this chapter that show where this incident took place.

9. M/Sgt Bledsoe did, in fact, survive the mortar blast. He was awarded the Bronze and Silver Stars for Valor, plus the Purple Heart for wounds he received in this incident.

10. This campaign was dubbed Operation Queen, a coordinated attack launched, as Ralph noted in his diary, at 11:00 am on November 16. The operation involved, in order from north to south in the Hurtgen, the 9th, 1st, 4th and 8th Infantry Divisions. The 8th Division operated parallel to and on the southern flank of the 12th Infantry Regiment, which is why Ralph refers to it in his memoirs as "our companion outfit in the forest." Queen's mission was to drive the Germans eastward out of the forest once and for all, thereby opening the way to the Cologne plain and the capture of the dams on the Roer River.

11. "In nine days... the 12th Regiment suffered more than 1,600 battle and non-battle casualties. Companies attacking the enemy often found themselves surrounded by other enemy forces that had infiltrated through the thick forest undetected. A regimental commander was relieved. [Opposing them was the] 275th Division ... bolstered by some 37 different German units" ("Bloodletting in the Hurtgen Forest," by Nathan N. Prefer in warfarehistorynetwork.com). Before being relieved on 8 December, the Fourth Division suffered over 6,000 battle and non-battle casualties. Casualties were so horrific and the Forest so dense and impenetrable that hundreds of bodies, both Allied and German, were never recovered. To this day, soldiers' remains continue to emerge from the dark and bloody ground of this cursed forest and are given their long-overdue honor and a decent interment (https://www.liberationroute.com/themed-routes/5/the-battle-of-the-hurtgen-forest). A fascinating and definitive history of the efforts on the part of the Army after the War to find, identify and re-inter Allied remains from the Hurtgen is available at https://www.dpaa.mil/Portals/85/Briefing%20Video%20Files/Huertgen_Project_Slides.pdf?ver=PEN_4b_yW2sEEUVTOYpdAA%3D%3D.

hot shards of metal and splintered, razor-sharp tree branches. We dug the log-covered holes to shield ourselves — often in vain — from the deadly blasts.

Equally terrifying were booby-trapped foxholes. As the Germans retreated, they left explosives in their abandoned holes that detonated whenever any of our guys dove in them as they pursued the retreating enemy. I recall one instance when we suddenly came under heavy artillery fire. We all dove under whatever cover we could find, and one poor soul pitched himself into one of the Krauts' booby-trapped holes. The explosion blew him back out of the hole — he was horribly wounded, but, pinned down by the shells raining down on us, we couldn't get to him. Helpless to do anything else, we alternated between cursing the Krauts and shouting to our comrade to hang on. We listened to him scream in agony for hours until he finally, mercifully, expired.

It was a weary, battered, depleted Fourth Infantry Division that was finally relieved by the 83rd Division at the edge of the Hurtgen Forest just on the verge of breaking out into the great plains leading to Cologne, Germany. Moving in a long convoy, we cleared the Forest on December 8, 1944, moving south.

We didn't know where we were going, but we were only too happy that it was out of that Forest. Every company was down to a handful of men. We lost thousands of men via the tree-burst route, by pneumonia, frozen hands and feet, mines, and of course small arms wounds. No longer, we thought, would we have to flounder in knee-deep snow, face the terrible German 88's, or live like animals in the frozen earth.[12]

Despite all the hardships it endured, the 12th Infantry Regiment acquitted itself well during its month in this terrible forest, though at awful cost. Our guys' efforts and sacrifice earned us a special "attaboy" from our Division Commanding General, Raymond O. "Tubby" Barton. General Barton's commendation was published the next month in the regimental newsletter, the *Big Picture*. It read as follows:

1. *At the close of the highly significant Hurtgen Forest operation, I wish to express to you and through you, to the officers and men of the 12th Infantry Regiment, my profound appreciation of the outstanding performance of your regiment during this series of engagements.*

2. *On 8 Nov 44, the 12th Infantry, following a long series of brilliant successes since the D-Day landing in Normandy, opened an offensive against the enemy in the region of Zweifall, Germany, the Regi-*

---

12. In one of the many ironies of the war, the 83rd and the 4th simply exchanged places: the 83rd took the place of the 4th in the terrible Forest, and the 4th took the place of the 83rd between Luxembourg City and the Saur River — a quiet sector where the battered division could rest and recover, and badly depleted equipment and materiel could be repaired, refitted, or replaced. This welcome interlude would last all of eight days.

ment was assigned the extremely difficult and vital mission of clearing a portion of the Hurtgen Forest of powerful enemy forces and fighting its way to the Cologne Plain.

3. *Throughout the entire campaign, the progress of the regiment was seriously hampered by an unusual combination of impediments created by weather and terrain. Unseasonal precipitation and damp, penetrating cold were a constant detriment to the health and well-being of the personnel, rendering their day-by-day existence well-nigh unbearable. The terrain was characterized by densely forested hills, swollen streams, and deep, adhesive mud, which retarded all movement of troops and vehicles.*

4. *Fully cognizant of the decided strategic advantage which this area afforded for effective defense, the enemy had prepared an elaborate system of mutually supporting fortifications, interspersed with ingeniously contrived obstacles of many types. Extensive mine fields and well-placed booby traps, in particular, exacted a heavy toll of casualties during the Advance. The effectiveness of enemy artillery and mortar fire was considerably enhanced by the frequency of tree-bursts in this heavily timbered area. Inasmuch as natural conditions precluded the employment of adequate aerial and motorized support, the burden of neutralizing fanatically defended enemy fortifications fell heavily upon the shoulders of the foot soldiers. Individual daring and initiative played a vital part in reducing each of these fortifications. The success of this campaign might well be attributed to the cumulation of countless acts of heroism by individuals and by groups of men fighting as well coordinated teams.*

5. *Problems of supply, personnel replacement, and evacuation of casualties were extremely complicated by mud, snow, and intense enemy artillery fire, both in forward and rear areas. Litter cases were frequently carried 1,000 to 1,500 yards in reaching points accessible to medical vehicles, and vast quantities of vital supplies were hand-carried hundreds of yards to forward positions.*

6. *Against one of the most formidable combinations of adversities ever faced by a military unit, the 12th Infantry pressed forward, day after day, inflicting severe casualties upon the enemy. The hard-won advances, although measured in yards rather than miles, possessed high tactical value. Under living conditions which alone called for heroism and physical stamina of a high degree, the 12th Infantry fought as a cohesive unit and carried the fight aggressively to the enemy. Each advance would have been impossible without the determination and cooperative fighting spirit displayed by the regiment. The battle casualties suffered and the loss of men from non-battle causes are the mark of the stern resistance and physical hardships overcome, and it is a tribute to the gallantry of the men of the 12th Infantry that losses in no way deterred their aggressive action. Although forced by the exigencies of combat to operate, upon many occasions, considerably understrength and without rest, the regiment, by their brilliant performance kept faith with its honored dead.*

7. *I commend your officers and men for the successful completion of a mission vital to the Allied cause. The deeds of the 12th Infantry Regiment shall not be forgotten as long as bravery and valor are honored and respected.*

(Signed). The Commanding General

Above: Major General Raymond O. "Tubby" Barton, Commanding General of the Fourth Infantry Division.

Below: This came out just after our part of the Battle of the Bulge ended December 24. Nice words from the Boss about the hard fight in the Hurtgen, right on the heels of another hard fight saving Radio Luxembourg and Army Headquarters in Luxembourg City. That last effort would earn us a Presidential Unit Citation. Shame of it is, the real heroes didn't live to read either one.

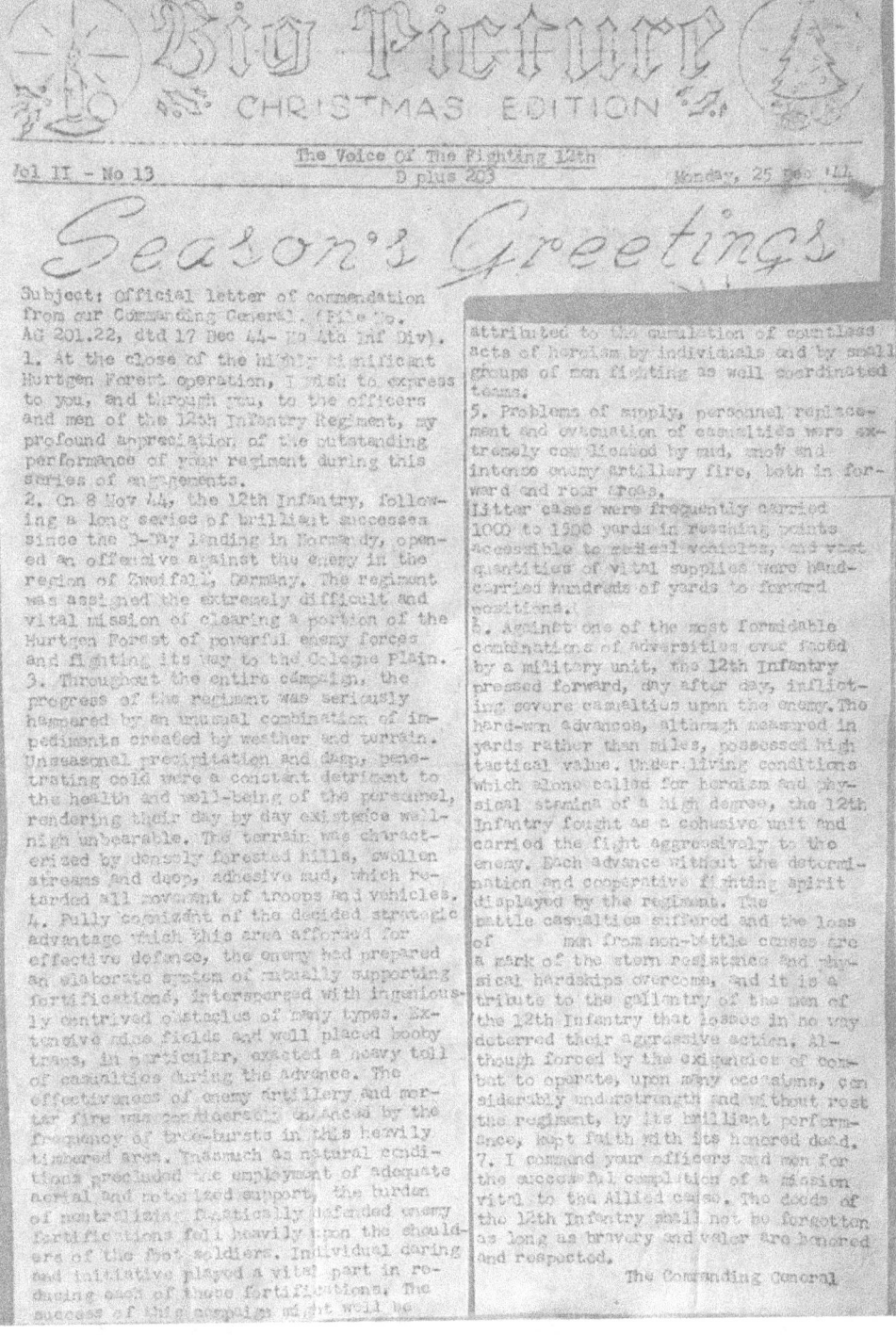

WARWICK at the front. Sgt. Joseph Sandoval, Gallup, N. Mex., 99th Div. artilleryman, reads Stars and Stripes with Warwick section during Ardennes lull. Pal is Pvt. Henry Schroeder, Wagner, S. Dak.

TWO-IN-ONE or battle-handling (top) demonstrated by Sgt. Richard Morgan, Youngstown, O. and S Sgt. James H. Puckett, Elgin, Ill. T/5 R. Fenk, Bliss, N.Y. (bottom) toasts K ration cracker with lighter.

Above, below right: Pretty typical scenes from the Hurtgen Forest that I saw and clipped from *Yank* and *S&S*.

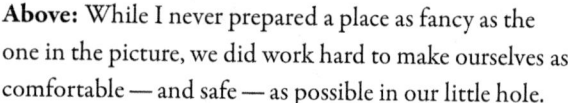

Above: While I never prepared a place as fancy as the one in the picture, we did work hard to make ourselves as comfortable — and safe — as possible in our little hole.

Up Front With Mauldin

"Them wuz his exact words—'I envy th' way you dogfaces git first pick o' wimmin an' likker in town.'"

Left: From my copy of *S&S* — My thoughts exactly on my first day on the combat front line in Hurtgen!

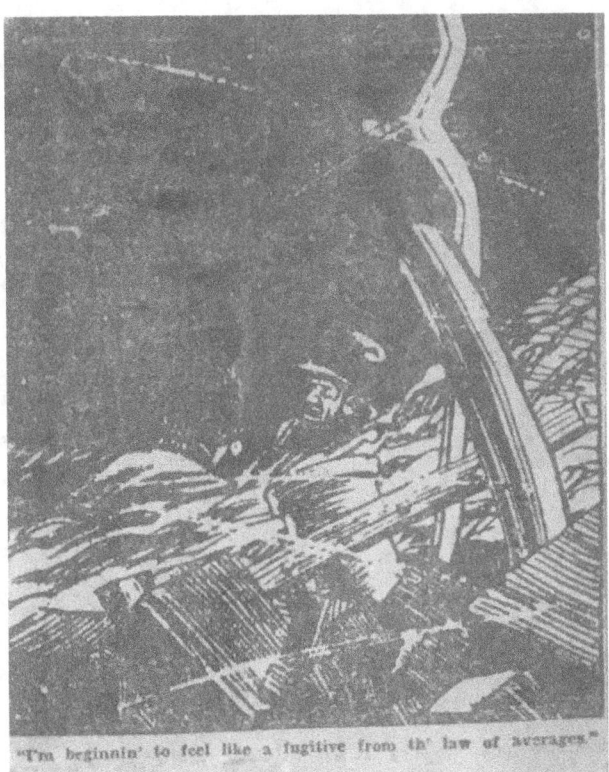

"I'm beginnin' to feel like a fugitive from th' law of averages."

Above: After surviving a terrifying night clutching my SCR-300 (**left**) and huddling as close as possible to a captured German pillbox while Kraut artillery streaked through the sky, I could fully sympathize with Willie!

Right: Were our letters from home ever read? Yes — at least by the censor in regimental headquarters — and, I feel certain, by others therein when there was anything worth reading.

Below and next page: Some of the many articles appearing in *Stars and Stripes* about the 4th Division during the Hurtgen Forest fight.

Fourth's 20,000th PW

A hungry-looking 30-year-old Austrian farmhand was the 20,000th prisoner captured since D-Day by the Fourth Inf. Div. A veteran of the Russian campaign, he told Lt. George E. Horn, of New York, that the fighting in the Hurtgen Forest was tougher than on the Eastern front.

Nazi Digs Own Grave With Wagging Tongue

WITH THE FOURTH INF. DIV.—1/Lt. William C. Staudt, of San Antonio, forward observer for a FA battalion, was adjusting fire by sound in the Hurtgen Forest, since the thick woods cut off visibility. He heard a shell land out in front, and a voice called out in German.

"....as that near you, Ludwig?"
"....o" said Ludwig, "it was 200 ....s to the right."

Staudt changed the range of the next round. Later that day, Ludwig's remains were picked up around the area.

....eaches His Limit

....fc Troy L. Doss is a patient ma.... was digging a foxhole when ....llet nicked his trousers leg. ....pt on digging. A second sh.... ....zed his legging. Doss kept ....gging. A third shot whizzed pa.... s finger. He stopped long enou.... see he wasn't hit and went ....gging. Then a fourth shot to.... rough his overcoat.

The Fourth Infantry Divisio.... ....ldier from Warrior, Ala., decid.... move. Foxhole? Yep, it was n.... ep enough.

....ut of the Frying Pan

T/5 Otho Ridgley, of 4th Infant.... ....vision's 70th Tank Battali.... ....ows what it means to hop o.... the frying pan into the fi.... ....gley was putting his shoe ....en a 240mm shell burst close ....shing off the heel of the sh.... ....dgley was sighing in relief, wh.... ....heard an ominous crack. T.... ....ne shell had splintered a t.... ....nk, and a pine crashed dow.... ....aking his ankle.

Nazi Apology

WITH THE FOURTH INFA.... ....TRY DIVISION, GERMANY, Nov. .... —German troops, who fired on .... medical jeep, sent a messenger fo.... ward with a white flag of truce an.... ....n apology for their action. Th.... ....ote also advised the Yanks the.... ....vould be unharmed if they wer.... ....orward to remove the bodies o.... ....he casualties and recover the.... ....eep.

Against the advice of some of h.... ....uddies, who suspected a trick, L.... ....loyd R. Golby, of South Orange ....J., removed his weapons an.... ....alked back with the Jerry mes.... ....enger.

The German helped Golby plac.... ....he first body in the jeep and cove....

....up. Then the two rode abou.... ....00 yards to the other one. Th.... ....erry then returned to the Amer.... ....an lines with the white flag t.... ....scort Golby back safely befor.... ....turning to his own lines.

Golby was a tech sergeant o.... ....Day when he landed with libera.... ....on forces. He received a battl.... ....eld commission and is now a bat.... ....llion S-4.

Nazi's Diary Reveals Might of Yank Assault

American soldiers of the First U.S. Army got an idea how it feels to be under one of their own attacks when they captured a German medic who had kept a diary of the fighting in the Gey-Grosshau sector. His story starts Nov. 26:

"The hours pass slowly and as I peer out of my hole the first dim light shows in the east. The is approaching. We expe. 'Ami' (German slang for American) to attack at 730. Then our fate will be decided. Our CO is looking toward Grosshau, he has a good vantage point.

"It's Sunday. My God, today is Sunday.

"To our left, machine-guns begin to chatter—and here come the Ami. In broad waves you can see him come across the field. Tanks all around him are firing wildly. Now the American artillery ceases and the tank guns are firing like mad. Can't stick my head out of my hole—finally there are three German assault guns. With few shots we can see several tanks burning once again. Long smoke columns are rising toward heaven. The infantry attack slows down— it's stopped!

Nazis Counter-attack

"Unbelievably, with this handful of men, we hold out against such attacks. And now we go forward to the counter-attack. The captain is leading it himself. Can't go far though, our people are dropping like tired flies. We have to go back and leave a number of dead and wounded.

"Nov. 27.—Amidst yells, they are breaking out of the forest again among their tanks. The Ami is getting closer now but the murderous fire of our MGs force him to the ground.

"Nov. 28 — Suddenly the tanks and then hordes of Amis are breaking out of the woods.

Take Up New Positions

"As they get within 70 paces, I turn and walk away. Very calmly, with my hands in my pockets. They are not shooting at me, perhaps because of the red cross on my back.

"Maybe I'll get out of this alive. If so, I can tell the story myself. But if I remain in this torn-up woods, perhaps a comrade will find this book and send it to my wife." Then the Germans surrendered.

Heroes of Hurtgen Forest Quite Willing to Forget It

*

WITH EIGHTH INF. DIV., GERMANY.—"The battle of Hurtgen Forest was supercharged hell!"

That's how 1/Lt. Jack R. Melton, of Co. I of the 121st Inf. Reg., summed up the campaign for Hurtgen Forest, west of the Roer River.

"The whole damn place was alive with mines," Melton recalled. "The Kraut artillery and mortar were the heaviest I've ever ducked. And to top it off, the weather was just what Adolf wanted—two weeks of rain and snow."

The tall Texan from Dallas told how the company jumped off one rainy dawn:

"We got only a few hundred yards when we ran into mines. Snipers fired from the edge of the woods. Then it came—the kitchen sink and all. Jerry threw in 120 mortars and 150 artillery. Everyone reserved himself a nice chunk of crater hole.

"It was the damnedest feeling of helplessness. We couldn't move ahead without a mine popping off or drawing mortar fire. The engineers made several attempts to get through but were unsuccessful.

Flushed Out Jerries

"Finally a patrol led by Lt. Stanley Schwartz, S/Sgt. Johnny Minik, Pvt. Harold Trusty and Pvt. Jessie Stevens pulled the company through. It was one helluva series of hand-to-hand battles.

"The Germans were always hidden in pillboxes or behind log bunkers. Artillery couldn't touch them, so our doughboys guided tanks to the bunker flanks and flushed out the Krauts by the hundreds."

Melton paused and reflected. "But it was not just the fellow firing the Garand who was the hero. Take those boys carrying supplies through that hellish forest for over 4,000 yards. It took our litter-bearers six hours to evacuate one casualty.

"And somebody should write a book on the work of the wiremen. Our battalion wire section laid over 40 miles of wire in an area extending over 3,100 yards.

"I'm with all my boys when they say history may remember this battle of Hurtgen Forest but we'd rather forget it."

*EIGHTH INFANTRY DIVISION WAS OUR COMPANION OUTFIT IN THE HURTGEN FOREST

"What makes you think you were run over by a Tank?"

Above: Our outfit suffered terribly during the fight in the Hurtgen — tree bursts from constant artillery and mortar bombardment, booby trapped concertina wire, minefields (not to mention pneumonia and trench foot and many cases of shell shock) took an awful toll...so much so that it was difficult to get on sick call. **Below:** Photo one of our guys took of one the dreaded German 88's.

Left: Red Warrior details carry the dead and wounded to the rear. The terrain in the sector assigned to the 12th Regiment in the Hurtgen was densely forested, hilly and, save for a few firebreaks, had no passable roads for vehicles to move men and supplies to the front lines or casualties to rear areas. As a result, a large of number of troops had to be diverted from the front lines to evac the dead and wounded. The Germans, knowing this, shelled the firebreaks constantly, adding to the bloody misery of the GI's in that cursed forest.

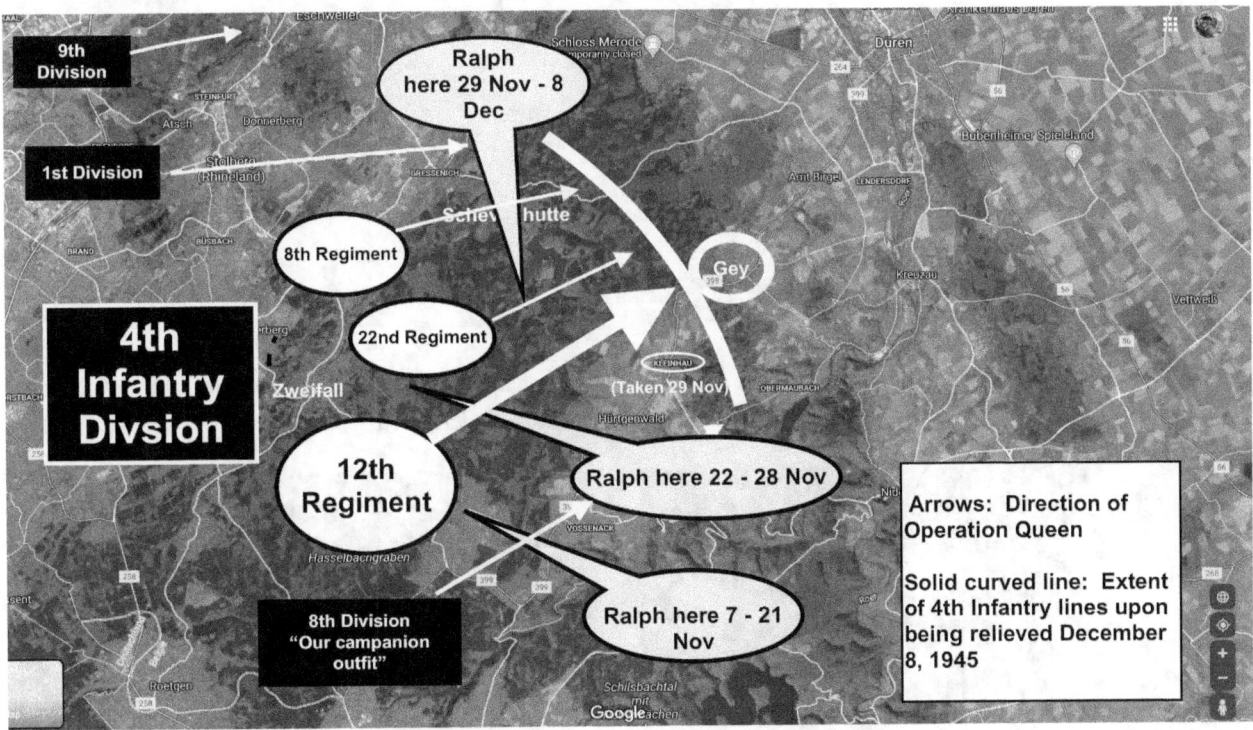

Above: In preparation for *Operation Queen*, the 121st Regiment of the 8th Infantry Division ("our companion outfit") entered the forest and relieved the battered southernmost companies of the 12th Regiment just prior to November 16. When *Queen* commenced, the 8th fought along a path roughly parallel to the southern flank of the 12th Regiment, capturing the village of Hurtgen and then continuing to push eastward, while the 12th Regiment pointed its battalions toward Gey, situated on the eastern edge of the Hurtgen. Of course, as a lowly Pfc, Ralph had no idea at the time of the involvement of the other divisions in *Queen* far to the north of his sector.

All three regiments of the 4th Division — 8th, 12th, and 22nd — reached their assigned objective after three weeks of desperate fighting; the Ivy Division was then relieved by the 83rd Division and exited the Hurtgen Forest on December 8.

Below: Sometime between 7 and 21 Nov: *"I found half the company hovering inside a captured Heinie pillbox… There was no room for me inside, so my SCR-300 kept me company outside… That night was a million years long."* The circle marks Rafflesbrand, from which Ralph "volunteered" to go to the bunkers at the front along the Weisser Wehe River.

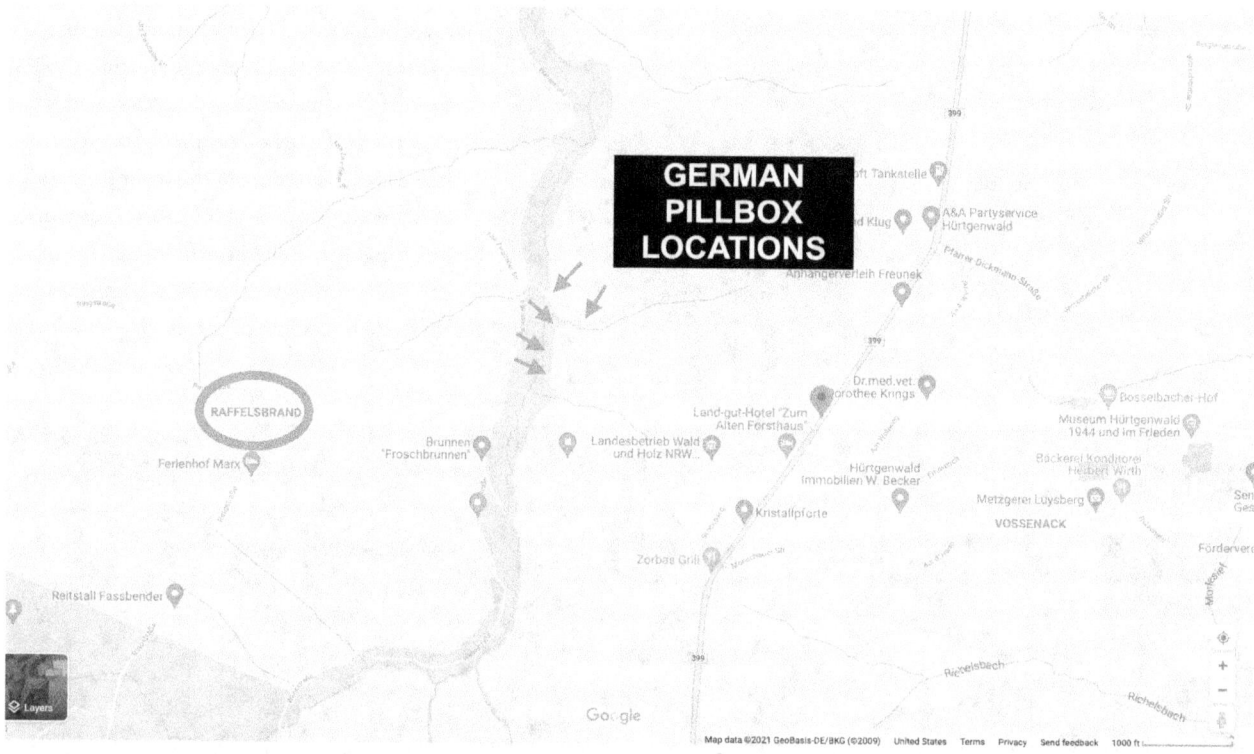

Below: Road used by jeep driver to transport Ralph to a forward position from Rafflesbrand, location of the 12th CP from 7-22 November 1944. Photos taken June 2018 with Ralph's children and grandchildren.

Rafflesbrand, Hurtgen Forest, Germany
12th Infantry Command Post
6 - 22 November 1944

Left: David Miles stands near the site of the 12th Regiment command post near Rafflesbrand, which was located in the background just beyond the tree line. During the war, this entire area was heavily forested.

Above: Satellite map showing Ralph's movements with the Regimental HQ company as *Operation Queen* pressed the Germans yard by bloody yard to the edge of the Hurtgen Forest. The three Ivy regiments reached the outskirts of Gey during the first week of December.

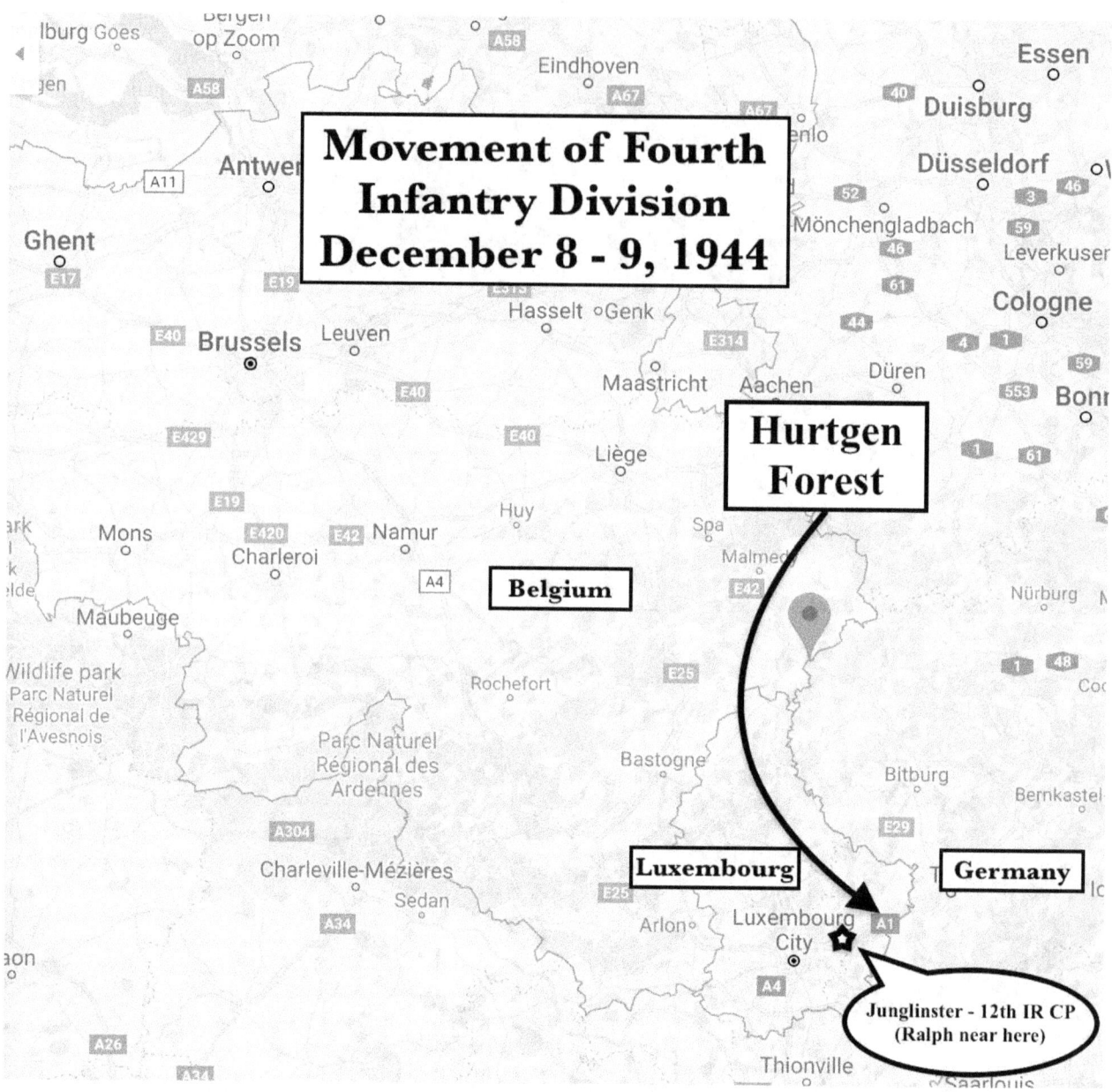

Movement of Fourth Infantry Division December 8 - 9, 1944

6. 'No Retrograde Movement': The Battle of the Bulge

"There will be no retrograde movement in the CT 12 sector."
— **Major General Raymond Barton, December 16, 1944**

Our convoy from the Hurtgen arrived in Luxembourg on December 9, 1944, and at our particular area, Junglinster, about dusk. Happily, we scrambled around and made ourselves reasonably comfortable in the houses, rounding up mattresses and firewood. It was snowing heavily and very cold. The people here spoke French or German or both. The 4th was supposed to take up a defensive position along the Luxembourg-German border east of the city of Luxembourg. Accordingly, the companies were deployed in the various small towns. The city of Luxembourg had only recently been taken and was considered a vital point in the general scheme of things. (These things I learned from reading material published much later, for at the time we all knew little about the "big picture" during the actual operation.)

It was here at Junglinster that I finally came to feel part of the outfit and no longer an outsider. Instead of being "volunteered" from pillar to post, I started working regular four-hour shifts on the regimental radio operated from a half-track.[1] Sgt. Nunzio Yocca was in charge of this part of the operation — a fine young man. The half-track was situated on top of a hill within sight of the giant tower transmitting plant of Radio Luxembourg, a highly critical point since this was the most important and powerful radio tower in Europe. It was kept in constant readiness to be blown up in case the Germans should get to where they might re-capture it.

Passes to the city of Luxembourg were passed out from the first, but it wasn't until December 16, 1944, that I finally got one. We drove to the city in an open truck on that day. It was about twenty miles, and the temperature was around ten above zero. I wore no topcoat because I wanted to buy some presents to send home and didn't want to have to carry them and the coat too, so I nearly froze for my foolishness.

---

1. Klinek observes that "some veterans looked unfavorably on untested men who were to take the place of their fallen buddies. Many experienced men worried that these green soldiers might not only get themselves killed but the veterans around them as well" (Klinek, pp. 136-137). Based on his comments here, Ralph appears to have passed his test while in the Hurtgen.

When I left for Luxembourg, little did I think the Germans would stage their final big counterattack that same day.

The City of Luxembourg was hardly hit at all by the war (so far) and the stores seemed plentifully supplied.[2]

My first move was into the ice cream shop, naturally, my first since England. When I had taken on considerably more ice cream than was good for me, I wandered about town looking at the store window displays and the huge placards bearing the likeness of the country's king and queen.

At one place I bought two hand-painted cards which I later fashioned into a birthday card for Sharon and an anniversary card for Eleanor. Making this purchase was most unique. After I had struggled with my English-German dictionary for some time, the saleslady said, "Just tell me what you want — I speak English!" I found this to be true to a large degree, much to my surprise, throughout Europe.

Partly to rest my aching feet from all the walking and partly to keep from freezing to death, my next purchase was a ticket to a movie. It was an American picture but the local language had been superimposed on the sound tract so I could only guess at some of it.

A month later I was destined to return on another day-pass to Luxembourg City with my pal Bob Weltzein. On this first trip I had no buddy and consequently did not enjoy myself as I did later.

Upon returning to Junglinster, things were in a turmoil. The last convulsive effort of the Nazis to stem the tide of defeat had begun that very day, December 16, 1944. This great battle later became known as the "Battle of the Bulge." Luxembourg City and Radio Luxembourg were prime objectives of the Germans, and we were directly in their path.[3]

With our companies so terribly undermanned from the Hurtgen fight, we knew it would be nip and tuck. But we held the southern shoulder of the "bulge." As our companies were hard pressed in their respective towns (Company E was completely cut off at Echternacht), three of us "volunteered" to take the half-track up to the front to try to make contact with the isolated men and to act as a relay station be-

---

2. The Germans had fled Luxembourg City without firing a shot at advancing American forces, which entered and liberated the city on 10 September 1944, some five weeks before Ralph's arrival in mid-December.

3. The Germans were aware of the weakened state of the Fourth Division battalions deployed along the Saur River, particularly those of the 12th IR. Fresh from the horrors of the Hurtgen Forest, the men were exhausted physically and emotionally, and their weaponry, including armor, badly needed repair and replacement. They had been sent to this sector because it had been quiet for some time and was considered a place where the battle-weary GI's could get some much-needed R&R, receive replacement troops, and get their equipment back in battle-ready condition.

tween the front and the 12th Infantry Regimental Headquarters.[4] All wires were being cut as fast as the stringers could run them, so all communications were via radio. Naturally, we were plenty busy.[5]

We sought out the highest points possible (and also the most exposed) for the half-track.[6] From our vantage points we could actually see the enemy troops as they advanced. The thin steel sides of the half-track seemed pitifully inadequate to us as we hovered there one-half frozen, and the other half scared to death. We covered the windshield and open areas of the half-track with several layers of blanket to keep out the light by which we operated the radio at night and to hold in what little body heat we could generate to keep warm.

We did contact some of the line companies although we never could make contact with Easy (E) Company, and we later learned that they had finally run out of ammo and supplies and had surrendered — that is, the few left.[7]

Prior to this battle, the Fourth Infantry Division was a part of the First Army under General Hodge, but we were now in Patton's Third Army.

Enemy pressure on our sector of the "bulge" was finally reduced just before Christmas, 1944, and the Fourth Infantry Division began to shift back into reserve, being relieved by the Fifth Infantry Division.

---

4. The main German offensive commenced in earnest at noon, but German infiltrators that morning had already crippled Allied comms, cutting wires and overrunning outlying posts. To make matters worse, many of the regiment's radios were being repaired and refitted, and its companies were widely scattered along the river. As if all this weren't enough, radio signals were hindered by heavily forested hills in this area, known as the "Little Switzerland" of Luxembourg. All of which explains Ralph's observation that he and his two half-track wireless radio buddies "were plenty busy" moving from hill to hill trying to establish wireless radio communication between the five front-line companies and the regimental command post in Junglinster. Things were so bad that, even though the "volunteer" half-track group was perched right above Echternacht on Hill 249, they were still unable to contact Company E in that village, where the beleaguered doughs made a heroic stand against the Nazis before surrendering to overwhelming odds around mid-day December 21 (Major Glenn Zargar, *Military Monograph: Defense of Little Switzerland* (May 1,1949 pp. 21-22).

5. The objective of the Germans' 212th Volksgrenadiers was to take Radio Luxembourg and Luxembourg City, which was a major military supply center, as well as the Headquarters of the 12th Army Group. In light of the critical importance of this area of the American lines, 4th ID CO General Barton issued his famous order, "There shall be no retrograde movement in the CT 12 sector." The men of the 12th rose to the demand of their commander, earning accolades from Barton and even from Patton himself. (Action Against the Enemy, Reports After/After Action Reports for the month of December, 1944. dated 6 January 1945; Fourth ID at the Battle of the Bulge, by Bob Babcock, FB post 16 December 2020.)

6. It was no coincidence that Ralph and his two half-track comrades were "volunteered" for this mission. In the words of 2nd Battalion Commander Glenn Zargar, "The nature of the terrain and the climate were such that we could depend on the 300 [SCR-300, which the radio half-tracks utilized], when no hill masses intervened" (Zargar, p. 42).

7. Taking to heart the command that there "be no retrograde movement in this sector," Company E carved for itself a legendary niche in the annals of US military history, fending off repeated German attacks, refusing to allow itself to be relieved or to take advantage of opportunities to withdraw. The beleaguered survivors of the company eventually yielded to overwhelming odds, surrendering early in the afternoon of December 21 (Johnson, p. 251).

I very happily remember receiving Eleanor's very first package on Christmas Day and a joyous occasion it was.

At this time, the Fourth shifted around in Luxembourg several times, getting ready to move up toward Bastogne to help reverse the "bulge" and drive the Germans back into Germany.

For its actions during this battle, our outfit won high praise from the higher ups, including a tribute from our divisional commander. The "vital installations" mentioned in his commendation include Radio Luxembourg. Also, in this citation the battle is referred to as the Battle of Luxembourg. I have previously used the phrase "Battle of the Bulge," by which it later came to be known in news reports. In this desperate action, cooks, bakers, mechanics, and radio operators, as well as many other technical personnel, were pressed into the fight to save Radio Luxembourg, and we "part-time" soldiers acquitted ourselves well, as witness the Presidential Unit Citation awarded the 12th Infantry Regiment, which follows next. At the end of the war, this Citation, combined with another Presidential Citation awarded the Division for its role in the liberation of Belgium, entitled all Ivy Men to wear the Belgian Fourragere Braid as well as the Unit Citation blue pin on our uniforms. [8]

The Citation was published three weeks after the battle in the January 10, 1944, edition of our Divisional paper, *The Big Picture*. I have highlighted in this transcription the section regarding the courageous doughs of Company E that I was never able to get on the line during the battle:

1. On 7 December 1944, the 4th Infantry Division completed with brilliance a month of bitter fighting in the Hurtgen Forest. On 9 December, ranks gravely depleted and weary, you went into a defensive position on a ten-mile front along the Sauer River, east and north of the City of Luxembourg holding the north sector of the front of the division. In this supposedly quiet sector, it was intended that you receive and integrate vitally needed reinforcements, renew equipment, and raise your level of supply. Procedures had been initiated which would have enabled you to effect reorganization, while at the same time securing a respite, well-earned and greatly needed, from offensive combat. This, however, was not to be,

2. On 16 December, the German Army launched a general offensive along the greater portion of the Western Front. With fury born of desperation, the assault was implemented to the limit of the enemy's resources. Penetrations to substantial depth quickly were made in areas to the north of your position and it was apparent the intended southern shoulder of the German offensive fell squarely within the sector of your regiment. Inflamed by his success to the north, the Germans drove to

8. Thanks to Michael Belis for providing valuable research in a FB post April 9, 2021 on the National 4th Infantry Division FB page. The doughs received their blue pins on board the *Sea Bass* shortly after sailing from LeHavre, France in July of 1945. The Presidential Distinguished Unit Citation itself was presented to the guys upon their arrival at Camp Shanks, New York on July 12, 1945, by the Post Commander, acting as representative for the War Department (Johnson, pp. 365 — 366).

establish and expand through you to the south. The full strength of the 212th Volksgrenadier Division was hurled against you. These were young, well-equipped troops, who, it had been learned, trained expressly for this gamble. Against them stood the 12th Regiment of United States Infantry, understrength and bearing scars of recent battle. At stake was the dominating ground and road net which would open to the enemy the City of Luxembourg, Radio Luxembourg, and military installations of prime importance, together with tremendous supply establishments.

3. At 1415 hours on 16 December, conscious of the imperative necessity that the initial German drive be held, this headquarters issued an order in substance as follows: "There shall be no retrograde movement in the 12th Infantry sector." The 12th Regiment held; held in the face of odds so ominous that it would be difficult even in retrospect, to believe possible had not one seen during months of continuous combat, the high courage and honor which marks all ranks of the 12th Infantry. Meeting the urgency of the operation and undismayed by the odds, your resistance never faltered, panic did not visit this field of battle, even with mass German infiltration as deep as four kilometers behind elements of your command.

4. *The history of America's participation in this war will be rife with acts of heroism, by individuals and by units, but, in the collective self-sacrifice and sheer courage, it is doubtful whether any shall surpass the stand made by Company E in Echternacht during this period. Isolated by the initial German thrust, this company was subjected, alternately, to enemy shelling and repeated attempts to storm the town. On 18 December, two friendly tanks contacted the company and urged an attempt to withdraw under tank protection. This the company refused, electing to continue its valiant defense, though hopelessly outnumbered. After long and costly operations, this isolated but determined group of American Infantrymen was overpowered, yet the tactical advantages accruing from Company E's gallant stand, and our admiration and debt for their complete subjugation of self-interest to a higher cause, cannot be overstated.*

5. Without exception each unit of your command exhibited a similar high resolve. So, too, was the spirit of the many units attached to and in support of your efforts during the course of the battle, although they would be first to resist any suggestion this had not been primarily a 12th Infantry show. The Commanding General, Third United States Army [Lt. General George S. Patton], characterized the Battle of Luxembourg as, in his opinion, "The most outstanding accomplishment of this division in its long series of engagements. The 12th Infantry Regiment is the unit meriting the greatest share of this high tribute. You held the Germans from thrusting south, you reduced the 121 Volksgrenadier Division to one-half its original strength, you denied him vital installations, seizure, and exploitation of which might have prolonged this war immeasurably and could have been regained only at a cost greatly in excess of that exacted in the defense; and equally

important, it was an extension of the line held by your regiment when relieved on 24 December, that reorganization of the allied position in the south was based."

6. More than once history has found its course shaped by the impact of the 12th Infantry, for yours is a regiment rich in deeds of courage.

The performance of the 12th Infantry Regiment in the Battle Luxembourg accords with the finest traditions of your enviable past, and I take this occasion to acknowledge and salute your magnificent service.

(Signed) The Commanding General, 4th Infantry Division

Transcription of the Distinguished Unit Citation Awarded the 12th Infantry its actions during the Battle of Luxembourg (the "Battle of the Bulge")

The 12th Infantry Regiment is cited for outstanding performance of duty in action against the enemy from 16 to 24 December 1944. With its weary ranks depleted following a month of bitter fighting in the Hurtgen Forest, the regiment moved on 9 December 1944 into a defensive position on a front extending ten miles along the Sauer River east of the city of Luxembourg, there to rest and await replacements.

At dawn on 16 December, the Germans launched a general offensive against the central portion of the Western Front, implementing the assault to the limit of their resources. The enemy quickly made substantial penetrations in areas north of the 12th Infantry's position, and, as it later became apparent, intended to place the southern shoulder of his offensive squarely within its sector, as he drove to expand south to match his seeming success in the north.

After intensive artillery preparations, which destroyed all wired communications, two regiments of the 212th Volksgrenadier Division were thrown across the Sauer River on 16 December, with the remainder of this division and one additional regiment, reinforced, following on the 17th. At stake was dominating ground and the road net which would open to the enemy the city of Luxembourg, with the 12th Army Group Headquarters, supply installations of great magnitude and Radio Luxembourg.

With crushing weight, the enemy swirled around the defenders, making infiltrations of battalion strength to depths of four kilometers. By nightfall of 16 December, substantial elements of six companies of the 12th Regiment had been surrounded or isolated. Yet, conscious of the imperative necessity of containing the initial German drive at all costs, the courageous units held firm.

On successive days the regiment was subjected to intensive shelling and enemy attempts to storm the villages and positions to which the determined defenders tenaciously clung. With great skill and fortitude, the resolute American infantrymen disputed villages, house by house, and ground, yard by yard, inflicting such fearful casualties on the enemy that by 22 December the strength of the 212th Volksgrenadier Division had been reduced by one half and its ability to operate offensively destroyed.

At no point had the regimental main line of resistance been pierced. Faithful to its rich tradition, with utter devotion to the task at hand the 12th Infantry met the critical urgency of the occasion and, undismayed by heavy odds, prevented the enemy from thrusting south, denied him vital military and political installations, seizure and exploitation of which would have had grave consequences, and held a position which enabled other units to batter the enemy flank when the heroic 12th was relieved on 24 December.

The courage and fighting determination of each member of the regiment in the stand along the Sauer River presented an inspiring example of the invincibility of free men at arms.

CG LAUDS 12th

Hq 4th Inf Div letter dated 4 Jan 1945, File No. AG 201.22, Subject: Commendation To: Commanding Officer, 12th Infantry.

1. On 7 Dec 1944, the 12th Infantry Regiment completed with brilliance a month of bitter fighting in the Hürtgen Forest. On 9 Dec, ranks gravely depleted and weary, you went into a defensive position on a ten mile front along the Sauer River, east and north of the City of Luxembourg holding the north sector of the front of this division. In this supposedly quiet sector it was intended that you receive and intregrate vitally needed reinforcements, renew equipment and raise your level of supply. Procedures had been initiated which would have enabled you to effect reorganization, while at the same time securing a respite, well earned and greatly needed, from offensive combat. This, however, was not to be.

2. On 16 Dec the Germans launched a general offensive along the greater portion of the Western Front. With fury born of desperation, the assault was implemented to the limit of the enemy's resources. Penetration to substantial depth quickly were made in areas to the north of your position, and it was apparent the intended southern shoulder of the German offensive fell squarely within the sector of your regiment. Inflamed by his successes to the north, the Germans drove to establish and expand through you to the south. The full strength of the 212 Volksgrenadier Division was hurled against you. These were young, well equipped troops, who, it had been learned trained expressly for this battle. Against them stood the 12th Regiment of United States Infantry, understrength and bearing scars of recent battle. At stake was the dominating ground and road net which would open to the enemy the City of Luxembourg, Radio Luxembourg and military installations of prime importance, together with tremendous supply establishments.

(Cont'd on column 3)

3. At 1415 hours on 16 December, conscious of the imperative necessity that the initial German drive be held, this headquarters issued an order in substance as follows: "There will be no retrograde movement in the 12th Infantry sector." The 12th Infantry held; held in the face of odds so ominous that it would be difficult, even in retrospect, to believe possible had not one seen, during months of continuous combat, the high courage and honor which marks all ranks of the 12th Infantry. Meeting the urgency of the occasion and undismayed by the odds, your resistance never faltered, panic did not visit this field of battle, even with mass German infiltration as deep as four kilometers behind elements of your command.

4. The history of America's participation in this war will be rife with acts of heroism, by individuals and by units, but, in collective self-sacrifice and sheer courage, it is doubtful whether any shall surpass the stand made by Company E, in Echternach, during this period. Isolated by the initial German thrust, this company was subjected, alternately, to enemy shelling and repeated attempts to storm the town. On 18 December, two friendly tanks contacted the company and urged an attempt to withdraw under tank protection. This the company refused, electing to continue its valiant defense, though hopelessly outnumbered. After long and costly operations, this isolated but determined group of American Infantrymen was overpowered, yet the tactical advantages accruing from Company E's valiant stand, and our admiration and debt for their complete subjugation of self-interest to a higher cause, cannot be overstated.

5. Without exception, each unit of your command exhibited a similar high resolve. So too, was the spirit of the many units attached to and in support

(Cont'd on page 2)

NEXT PAGE

THE STARS AND STRIPES

Daily Newspaper of U.S. Armed Forces — in the European Theater of Operations

Ici On Parle Français
J'en voudrais un de plus.
John voodRAY von duh plew.
I'd like one more.

1 Fr.　　New York — PARIS — London　　1 Fr.　　Monday, Dec. 18, 1944

Nazis Smash Back Across Border; Luftwaffe Aids Drive on Hodges Line

97 Enemy Planes Shot Down Aiding German Drive

The Luftwaffe came out of hiding yesterday to hurl its biggest tactical force since D-Day at U.S. forces on the Western Front, the Ninth Air Force announced last night, but by nightfall 97 of the 450 Nazi planes encountered during the day had been destroyed by Ninth fighter-bombers.

Thunderbolts, Lightnings and Mustangs of the Ninth flew more than 1,150 sorties against the Luftwaffe and against Nazi counter-attacks on the First Army front, while continuing their blows behind the lines in the Third and Ninth Army sectors. Their 97-plane bag set a record for air action on the front. Thirty-one U.S. planes were reported missing.

Jumped by German Planes

Stars and Stripes Correspondent Julius Grad reported that the gigantic air battle, reminiscent of the dog fights over Britain during the '41 Blitz, started yesterday morning when between 200 and 250 German planes jumped Thunderbolt formations above the Saar area, just east of the Rhine and southwest of Cologne. In this fight alone, at least 21 FW190s and ME109s were shot down.

The Front as Counter-Attack Opened

Re-Enter Belgium, Luxembourg at Three Points

German troops, supported by tanks and the Luftwaffe, smashed at American First Army lines on a 70-mile front from Duren to Trier yesterday and drove back across the German border into Belgium and Luxembourg at three points along the invasion route the Nazis took in 1940.

Famed author Ernest Hemmingway travelled with the 4th ID during the war. I was never so fortunate as to meet the man myself. **Below** is an article from *S&S* about one incident that happened involving Hemingway, who actually took up arms to help fight off the Germans during the Bulge.

— Thanks, Mr. Hemingway —
Archie Gets Hep in a Hurry

By Joe Weston
Stars and Stripes Special Writer

WITH FOURTH INF. DIV.—Credit writer Ernest Hemingway with an assist, Pvt. Archie Pelky, of Canton, N. J., with a hit and the Nazis with an error. The box score follows:

When the Jerries recently crowded Luxembourg City so tight that Fourth Div. MPs, cooks and jeep drivers were sent to man the rifle companies on the line, Pvt. Pelky, jeep driver assigned by PRO to Hemingway, was one of those selected to become an over-night infantryman. Pelky had never actually seen combat.

He asked Hemingway for advice.

Pelky got orientated in an hurry.

"Fight like hell," said Hemingway, "dig a hole and be ready to take over if the lieutenant gets knocked off."

Next day, Pvt. Pelky hit the line. He was handed a bazooka and told to "find a tank." He had never fired a bazooka before but he learned damn fast when a Nazi medium started for his squad.

The second day found Pelky a machine-gunner. He'd never fired that either. He caught on fast.

The third day he replaced a wounded BAR-man. Needless to say, he hadn't ever fired a BAR either.

The Jerries were stopped cold and Luxembourg saved. But Pelky didn't go back to the wheel of Hemingway's jeep.

He got trench foot.

The 12th Infantry Regiment is cited for outstanding
performance of duty in action against the enemy from 16 to
24 December 1944. With its ranks depleted following
a month of bitter fighting in the Hurtgen Forest, the regiment moved on 9 December 1944 into a defensive position on
a front extending ten miles along the Sauer River east of
the city of Luxembourg, there to rest and await replacements.
At dawn on 16 December, the Germans launched a general offensive against the central portion of the Western Front, implementing the assault to the limit of their resources. The
enemy quickly made substantial penetrations in areas north of
the 12th Infantry's position, and, as it later became apparent,
intended to place the southern shoulder of his offensive
squarely within its sector, as he drove to expand south to
match his seeming success in the north. After intensive
artillery preparation, which destroyed all wire communications,
two regiments of the 212th Volksgrenadier Division were thrown
across the Sauer River on 16 December, with the remainder of
this division and one additional regiment, reinforced, following on the 17th. At stake was dominating ground and the road
net which would open to the enemy the city of Luxembourg, with
the 12th Army Group Headquarters, supply installations of
great magnitude and Radi Luxembourg. With crushing weight,
the enemy swirled around the defenders, making infiltrations
of battalion strength to depths of four kilometers. By nightfall of 16 December, substantial elements of six companies of
the 12th Infantry had been surrounded or isolated. Yet,
conscious of the imperative necessity of containing the initial
German drive at all costs, the courageous unit held firm. On
successive days the regiment was subjected to intensive shelling and enemy attempts to storm the villages and positions to
which the determined defenders tenaciously clung. With great
skill and fortitude, the resolute American infantrymen
disputed villages, house by house, and ground, yard by yard,
inflicting such fearful casualties on the enemy that by 22
December the strength of the 212th Volksgrenadier Division had
been reduced by one half and its ability to operate offensively destroyed. At no point had the regimental main line
of resistance been pierced. Faithful to its rich tradition,
with utter devotion to the task at hand, the 12th Infantry met
the critical urgency of the occasion and, undismayed by heavy
odds, prevented the enemy from thrusting south, denied him
vital military and political installations, seizure and exploitation of which would have had grave consequences, and held a
position which enabled other units to batter the enemy flank
when the heroic 12th was relieved on 24 December. The courage
and fighting determination of each member of the regiment in
the stand along the Sauer River presented an inspiring example
of the invincibility of free men at arms.

SEAL

THE ADJUTANT GENERAL'S OFFICE

OFFICIAL COPY

WAR DEPARTMENT
SEAL

OFFICIAL COPY OF PRESIDENTIAL CITATION ISSUED
TO THE 12TH INFANTRY REGIMENT, 4TH INF. DIVISION
(MY REGIMENT)

Below: First page of the January 8, 1945, *Life* magazine article about our Regiment's role in the Ardennes Offensive.

LIFE

8, No. 2   January 8, 1945

AMERICANS BATTLE THE GERMAN BIG PUSH

Is AND GENERALS FOUGHT HEROICALLY TO STEM ENEMY RUSH

by CHARLES CHRISTIAN WERTENBAKER

reacting to the great German counteroffensive the U. S. press and public again shown a deplorable lack of stability, blowing either too hot or too . When the Germans advanced, headlines often cried disaster. When they stopped, the headlines acclaimed victory for our side. Neither of these was true. This state of affairs is due partly to the U. S. love of big news, ther good or bad, and partly to a thick haze of military censorship, which it hard for people at home to get a complete or balanced picture of what ing on at the Western Front. To put the picture in proper focus LIFE its chief correspondent at the front to write a sober, balanced account of happened when the Germans broke through in Belgium and Luxembourg.

LUXEMBOURG

the morning of Saturday, Dec. 16, 1944 the situation of the 12th , S. Infantry Regiment, Colonel Robert H. Chance commanding, , as the commanding officer later put it in a laconic official report, "far n favorable." The 4th Division, of which the 12th was a part, had been ved after weeks of bitter fighting in the Hürtgen Forest (LIFE, Jan and sent to a quiet sector in Luxembourg to rest and rebuild its ngth. Its front ran for nearly 35 miles along the west bank of the er and Moselle Rivers and all three regiments were in the line. Be- se of its large sector and a shortage of equipment, communications e strained. Its artillery was scattered and shells were scarce. Its ached tank battalion, which had also taken heavy punishment in the rtgen Forest, was trying to repair its tanks in spite of an acute rtage of parts. One fourth of its tanks were stripped for cleaning. ny others would not run. The battalion had only 26 tanks which ld be considered operational. The 4th Division, in other words, was in condition for a fight.

Nobody expected it to have to fight. Along a 70-mile front between area of Monschau and the area of Echternach, American troops were tly spread. We were massing our strength in other sectors and the my was considered too weak to attack us. And so inside the small ns of eastern Luxembourg, in Echternach, Berdorf, Lauterborn and twin towns of Osweiler and Dickweiler, the officers and men of the h Infantry loafed and enjoyed the local beer. Outdoors, in the patches

of hardwood forest and in positions sheltered by small hills, they made themselves as comfortable as possible in rain that drizzled all day. During the night the enemy usually sent over a few shells and the regiment artil- lery wasted a few of its rations on him.

During the night of Dec. 15–16, the enemy behaved as usual. Then just after 6:30 a. m. he increased his artillery fire. Through the half- light shells whistled in and burst in all the pretty toy towns near the river. They were especially heavy in Echternach. Through spies or extraordinarily good observation the Germans had plotted the location of battalion and company command posts and these received heavy fire. By 9 o'clock the regiment had lost wire contact with all its units below battalion level. Just after 9 a. m. the enemy infantry attacked.

In fog that lay thick on the sides of the hills and blotted out the towns, rifles and machine guns spoke and were answered. A dozen small battles developed. But because of the fog and the surprise and because their wires were cut, the units of the regiment did not know what was happen- ing to the other units, nor did they know what they were supposed to do. And so what happened during the rest of that day and during the days which followed might be taken as an example of what good soldiers do when they are infiltrated and cut off and surrounded.

For a long time after the battle began Colonel Chance himself knew little of what was going on. The telephone in his command post had warned him at 9:21 to be alert; the division to the north was getting ac- tivity. By 9:27 he had passed on the warning in his foghorn voice to all his battalion commanders. At 9:45 Company F reported from Berdorf that it was being attacked by a 15-man patrol armed with automatic weapons. This patrol later proved to be an entire battalion. From Lauterborn Company G reported that it was being attacked by a squad. This later proved to be another battalion. The colonel ordered the light tanks guarding Radio Luxembourg near Junglinster to be ready to go to the aid of Berdorf and Lauterborn.

At 10:49 E company was surrounded in Echternach. At 11:37 Com- pany G was surrounded at Lauterborn. At 12:50 Company I was sur- rounded at Dickweiler. At 14:15 (2:15 p. m.) Major General Raymond Barton, commander of the 4th Division, issued the following order:

CONTINUED ON NEXT PAGE

On this and the following page are photographs that were captured from the Germans during the battle and were included in the *Life* magazine article about the fighting on the southern shoulder of the Bulge involving the 12th Regiment.

In captured German picture Nazi Tommy-gunner trudges along road littered with burning, abandoned U.S. vehicles.

Left and below: These photos provide graphic evidence of the degree of surprise the Germans achieved on December 16. Stunned and ill-prepared US troops reeled in the face of the initial German wave, abandoning vehicles and equipment before finally regaining their footing and blunting the onslaught. The exception to this was the valiant Red Warrior Regiment of the Ivy Division, whose doughs were determined that there would be "no retrograde movement in [their] sector."

Captured U. S. equipment, including a jeep, is inspected by German soldiers in Belgium. The mud under their feet is evidence of the planeproof bad weather, one of the greatest allies of the German offensive.

Right: The Germans caught many units before they could retreat, and in such overwhelming numbers that the doughs could not fight the Nazi's off. As a result, a number of units — like Company E of the 12th Regiment — did all they could before being overwhelmed and taken captive.

Column of U. S. prisoners being herded to rear passes German King Tiger tank. Prisoner toll was high in first few days.

Dead Americans are looted of equipment by two scavenging German soldiers. One man at left has been robbed of shoes.

Left: Photos such as this no doubt further intensified GI anger that was already burning fiercely following the Malmedy massacre, as well as a number of other Nazi atrocities committed during the Ardennes Offensive.

The fact is, by this time in the war scavenging was often a matter of sheer survival for the average German soldier, as the Wehrmacht was experiencing crippling shortages in supplies, manpower, fuel and matériel.

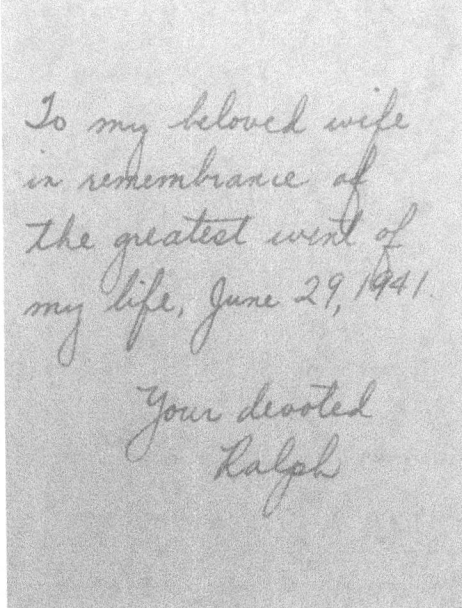

Above: The two hand-painted cards that I purchased in Luxembourg City 16 December. I practiced my calligraphic skills to fashion a birthday card for Sharron and an anniversary card for Eleanor.

WORTH 10¢

Left: As I moved from country to country, I picked up some souvenir samples of currency, coinage and stamps. Because Luxembourg was and is a bi-lingual nation, one side of the currency is in French while the other side is in German.

Left: Theatre tickets I bought in Luxembourg City. Starting at the top, it reads, "Capitol Cinema place card, Balcony, Seat 1." Dated 16 December 1944 — while I sat in a cozy theatre, bad things were happening back at the Front, as I would soon discover ….

STOOD RIGHT HERE SEVERAL MIN

Luxemburg
Gesamtansicht.

While in Luxembourg City, I bought some postcards, including one on which I drew an "X" that marks the spot where I stood on a street overlooking a deep valley on that bone-chilling day (**above**).

that I
the out-
Instead
lar to
4-hour
operated

While at Junglinster, I and my fellow squad members worked the regimental radio from a half-track just like the one pictured here (**left**). When the battle broke out on 16 December, we bounced frantically from hilltop to hilltop trying to contact the five front-line companies and relay their communications back to the CP in Junglinster (the Heinies had cut the wires, so all comms had to be wireless). It was a harrowing experience for sure. Had to skedaddle from Hill 249 as we saw German troops about to surround our position as they crossed the Sauer river below us.

GIs Hold Fire, Wipe Out Nazis

WITH FOURTH INF. DIV.—A small group of tankmen, engineers and tank destroyer men of the 4th Inf. Div. wiped out a Nazi company by waiting to see the whites of the German's eyes.

The Nazis were advancing along a draw in a "V" formation. The Yanks held their fire. As the enemy came within short range, Lt. Marvin C. Weber, of Lincoln, Neb., and the 70th Tank Bn., gave the order.

The Yanks mowed down every member of the German company except one man who was taken prisoner.

Fourth Div. infantrymen, crossing the battle scene the following day, counted the German dead. Their stiff, snow-covered bodies formed a crude "V"—but not for victory.

I kept a couple of wryly humorous articles I cut out of *Stars and Stripes* about the 4th Division during the Bulge. The first article (**left**) recounts how the 70th tank battalion dropped an approaching group of Krauts literally in their tracks by waiting until the last moment to open fire.

The second (**below**) tells the harrowing story of Sgt Doyle of the 22nd Infantry Regiment, who managed to escape a sure death by pretending to be dead — he endured being searched 19 times by passing Nazi soldiers for twelve hours ... twelve hours of horror that I can't imagine and am so grateful I never had to experience. Doyle was one mighty tough, brave dough!

Twelve Hours Face Down in the Snow
Pilfering Nazis Search 'Dead' GI 19 Times

WITH FOURTH INF. DIV.—When S/Sgt. Doyle C. Hopper of the 22nd Inf. Regt. heard footsteps coming his way, he knew they were German, decided to play dead and lay face down in the snow.

His heart beat wildly as two Germans rolled him over and slowly went through his pockets, but he managed to maintain his lifeless appearance. When the Nazis left, Hopper sighed with relief.

During 12 hours in that position, he was searched 19 times by souvenir-seeking parties. Several of them felt his pulse.

"I had my biggest scare," Hopper said later, "two big Krauts gave me the once-over. One of pulled out a knife and pointed it in the g_ direction of my throat, but I guess they figu_ wasn't necessary because they finally walked

Hopper sweated it out there on the snow_ nightfall, when he made his way safely back battalion CP.

I deemed the cartoon at **left** worthy of retention because those "ack-ack" guys (anti-aircraft) didn't in fact have many German airplanes to practice on (mostly due to the shortage of fuel), but the irony of the cartoon comes out in the statement, "never had it so good." These guys always had to emplace the weapons in wide-open fields, which left them exposed both to the elements and to enemy fire. Many were the slurs cast their way by the infantry "dog-faces," but (as this cartoon so well illustrates), theirs was such a miserable existence no dogface really envied them!

This "Hubert" cartoon was so typical of our squad that I cut it out of the *Stars and Stripes*. Kimmel was our section leader, a staff sergeant, who was forever running up with unpleasant news. Verga, the inveterate "fixer," can be seen working on the 50-calibre machine gun. Harris is in his typical position, his nose buried under the engine hood (he was the half-track driver). Weltzein, our only college graduate, was usually most meticulous about his personal appearance. And yours truly could typically be found either digging a slit trench (an Army "relief" project) or using one. Yocca was in charge of the half-track crew and T/4 Sergeant as well as a generally good handyman. Emery was a quiet and efficient young fellow from Erie, Pennsylvania.

The satellite map **below** shows the disposition of 12th Regiment rifle companies along the Saur River, which marked the border between Luxembourg and Germany. Only five companies — the 2nd Battalion, commanded by Major Glenn Zargar — were placed in villages astride the main transportation routes to the interior from the Sauer River — after all, this was supposed to be an inactive front, an ideal place to relax, recreate and refit — with the remainder of the regiment held in reserve. [US Army Center of Military History — *The Ardennes/The Battle of the* Bulge, "Chapter X: The German Southern Shoulder is Jammed" (https://www.history.army.mil/index.html)]. All the companies were exhausted and understrength, and their equipment was in poor repair from the Hurtgen struggle — and they sat directly in the path of the German 212th Volksgrenadier Division, whose troops crashed across the Saur River in force around mid-day on December 16, 1944.

(Map based on a sketch in Johnson, p. 232)

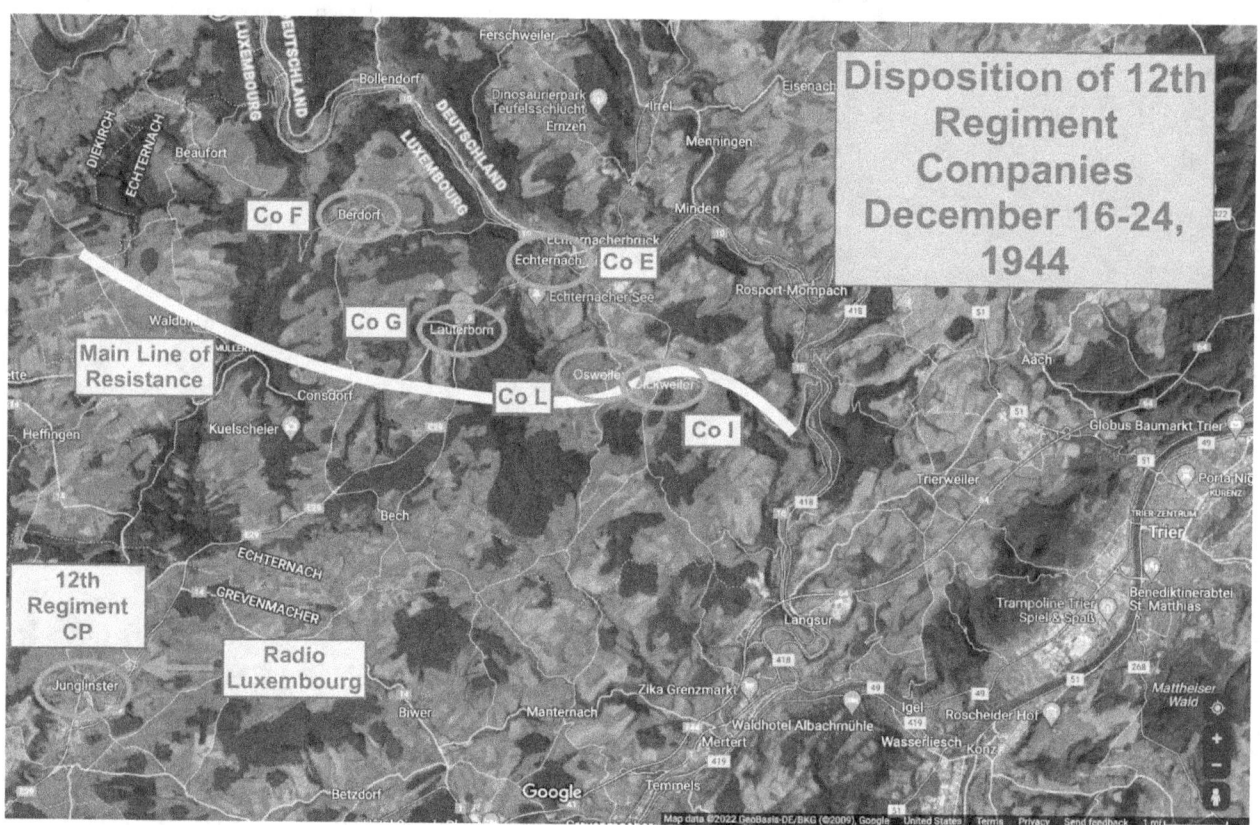

Hill 249 overlooking Echternacht

Outside Echternacht:
Ralph observed Germans crossing the Sauer River from Hill 249 in the background
(Picture taken June, 2018)

Left: *"We sought out the highest points possible (and also the most exposed) for the half-track. From our vantage points we could actually see the enemy troops as they advanced."*

Ralph's family poses about two miles west of Hill 249, one of the vantage points from which Nunzio Yocca's 3-man radio crew tried to contact the five rifle companies, including the ill-fated Company E in Echternacht below.

Right: That Ralph and his comrades could not raise Company E in Echternacht from atop Hill 249 (indicated by the circle), a high ridge immediately above the village, shows just how severe the communications issue was for the American commanders as they struggled frantically to coordinate resistance to the German onslaught. The lines on this military map indicate German army advances across the Sauer River and into the sector defended by the 12th Regiment.

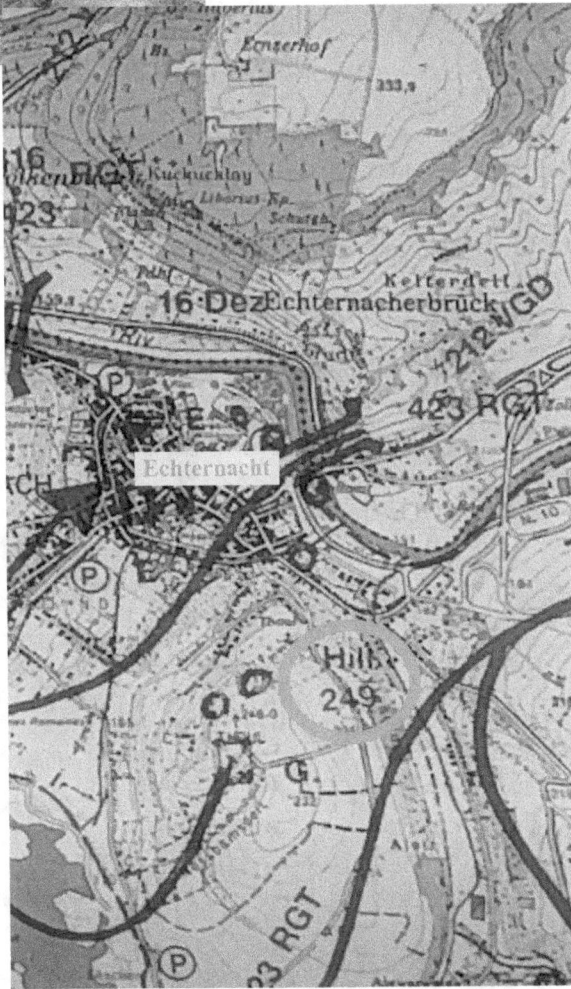

Above: Photo taken in 2023 from the brow of Hill 249. In the background, the village of Echternacht nestles at the foot of the hill on the banks of the Sauer River. The 212th Volksgrenadier Regiment splashed across the Sauer 16 December. Ralph, perched atop the hill in their regimental radio half-track could never raise Company E on his wireless, and had to evacuate quickly as they could see the waves of Germans swirling around the base of the hill.

Left: David Miles stands in a dirt road that leads to the top of Hill 249 on the opposite side from Echternacht — very possibly the same route followed by Ralph's halftrack radio squad in December, 1944.

Below: 2018 — Patrick, David and Nancy Miles outside the Company E command post building in Echternacht, Luxembourg. Ralph was never able to contact the surrounded Company during the battle. The memorial plaque on the wall of the building behind the trio reads:

"During the Battle of the Bulge, this place was heroically defended by soldiers of E-Comp, 12th Regt, 4th US Inf Div. Their sacrifice delayed the enemy advance and contributed to the final victory.

We shall remember."

Above: Ralph's family visits Radio Luxembourg in Junglinster in June of 2018. Pictured with them is their guide, Doug Mitchell (in the background in light colored shirt, second from left).

Luxembourg City - 19 June 2018

Above: Standing in the same spot where Ralph "stood for several minutes" while on a day pass in Luxembourg City 16 December 1944.

Small children, left to right: Eleanor Kate Miles, Adelyn Lee Miles, Piper Hannah Miles, Ansley Grace Miles.

Standing, left to right: Madison Yvonne Hendricks, Colin Gregory Hendricks, Emily Kathleen Miles, Nathan Lee Miles, Nancy Yvonne Miles, David Wayne Miles, Ashley Yvonne Miles Hendricks, Allan Miles Hendricks, Patrick David Miles, Jessica Raquel Miles.

Above: Ralph's *Veterans of the Battle of the Bulge* membership poster that he framed and displayed with quiet pride in his home in the latter years of his life.

"Your fight in the Hurtgen Forest was an epic of stark infantry combat; but, in my opinion, your most recent fight — from the 16th to the 26th of December — when with a depleted and tired division you halted the left shoulder of the German thrust into the American lines and saved the city of Luxembourg, together with the Twelfth Army Group and the tremendous supply establishments and the road nets in the vicinity — is the most outstanding accomplishment of yourself and your division."

— **Lt. General George S. Patton, Jr., commanding general, U.S. Third Army, to Major General Raymond O. Barton, commanding general, Fourth Infantry Division**

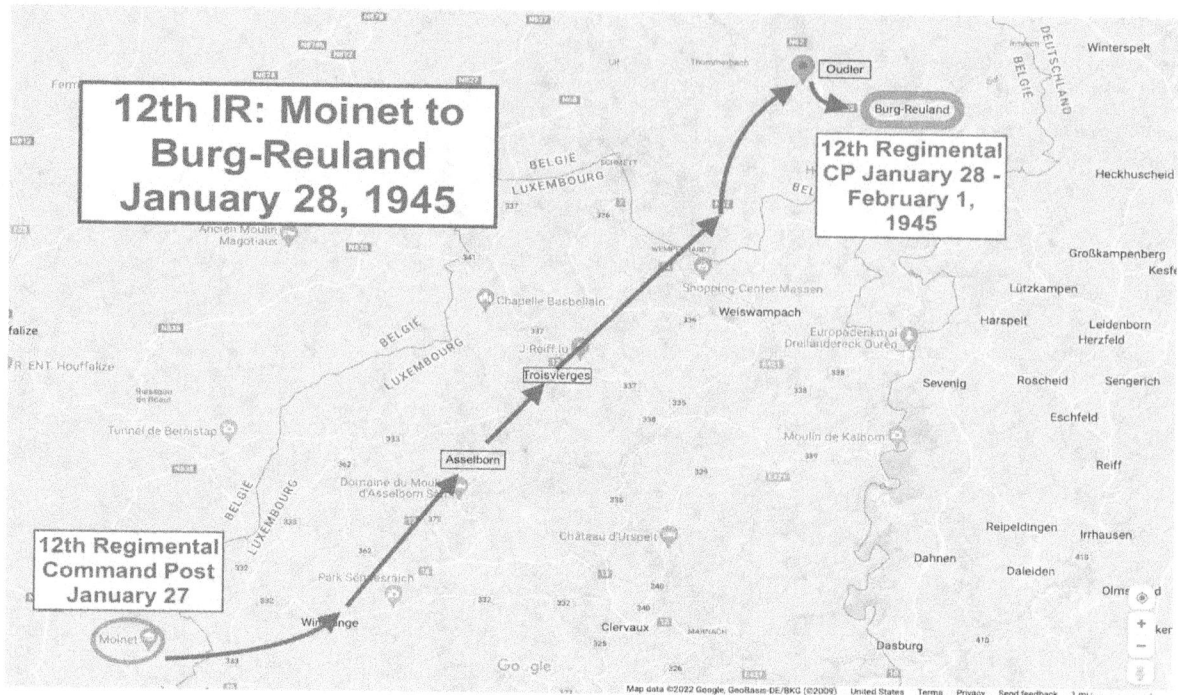

These maps show the progress of the Red Warrior Regiment from Moinet, Belgium, to Berg Reuland, just inside Belgian territory January 27 — February 1, 1945. From there, the Regiment crossed the Siegfried Line into Germany, encountering scattered resistance from the Wehrmacht. The 12th reached Buchet on February 8, 1945, where they received air-dropped supplies in preparation for the assault on Prum.

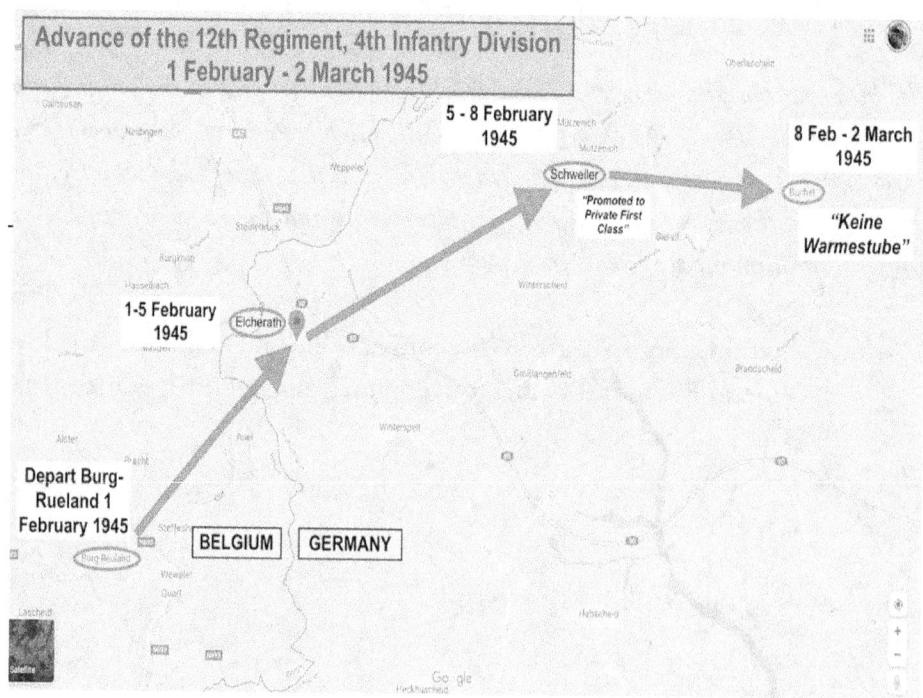

7. Hanging Our Wash on the Siegfried Line

We're going to hang out the washing on the Siegfried Line,
Have you any dirty washing, mother dear?
We're gonna hang out the washing on the Siegfried Line,
Why? 'Cause the washing day is here.
— **Jimmy Kennedy**

We got a rest at Hemstal, Luxemburg from December 28, 1944, to January 17, 1945. The Radio Section was billeted at the Café Schmit-Reuter, a nice place — no longer a café, of course. It was here at Hemstal that we were finally issued sleeping bags. While the snow continued to fall heavily, I spent the next several days in my spare time sewing my two blankets into or onto the inner wool sleeping bag, after which the entire affair was stuffed into the canvas outer cover. From that point on, sleeping was a cinch.

While at Hemstal, on January 15, 1945, Weltzein and I got a pass to Luxembourg City (his first, my second). At this time I had learned that Eleanor was pregnant, and I wanted so much to buy something nice to send to her, but much to our dismay, we could not buy any such items without ration stamps. So we quenched our disappointment in typical fashion: a pie and ice-cream eating contest. The principals — Weltzein in one chair at 190 pounds, Miles in the other at 170. You would have split your sides at that sight! The thermometer read ten degrees. There we sat — huddled in chairs with necks drawn far down past the upright collars of our overcoats, wool knit caps pulled well down over stinging red ears, topped by helmets, complete with glove-covered hands — practically numb — wolfing down pie and ice-cream! Perhaps Weltzein and I didn't look quite as shoddy as Hubert and friend (**previous page**) — but the girls surely looked fully as grand!

After a good rest in Hemstal, we began to jockey for position to re-enter Belgium, beginning January 17, 1945, at Meispelt. We did not spearhead the drive to relieve the troops surrounded at Bastogne, but we followed the front-line troops through the narrow corridor left by Patton's armored outfits.

At Gilsdorf, the last town I can remember before leaving Luxembourg, I was told I could now wear the Combat Infantryman's Badge, or CIB. The First Sergeant gave me the badge, which, in addition to the prestige, meant $10.00 per month "combat" pay.

* * * * *

Before leaving the Luxembourg theatre, the Regiment continued to battle German units frantically trying to secure safe passage across the Siegfried Line following the failed Ardennes offensive. Though the

American forces did considerable damage to the demoralized and tattered German remnants, many still managed to withdraw behind the West Wall.

Ralph and his half-track radio buddies departed from their post at Junglinster on December 27, 1945, and shifted from village to village as the front lines fluctuated in response to lingering German resistance throughout much of January.

Finally, the regiment departed Gilsdorf early on the morning of January 27 and opened a CP in Moinet, Belgium, at 11:30 am that same day. It was, as Ralph wrote, a very short stay - radio orders went out to all 12th Infantry companies at 9:20 pm to "be prepared for possible movement early tomorrow morning [January 28]. Details later."[1] The next day, oral orders were issued to move out via Asselborn, Troisvierges and Oudler, and the Moinet CP closed up shop by three o'clock that afternoon. The battle-hardened Red Warriors no doubt hoped the worst of the bloodletting had ended with their smashing victory in Luxembourg. They would soon find out otherwise.[2]

We pick up Ralph's account here, as he and his comrades follow behind Patton's Fourth Armored Division to a point outside Bastogne, and from there to Burg Rueland on the German border...

* * * * *

The Sergeant said this change in status (combat pay) would be retroactive to October 22, 1944, which meant I would be able to send some more money home to help Eleanor with the forthcoming hospital bills.[3] As best I can figure out, the date was January 22, 1945.[4] On the next page is a map tracking our movements after the Bulge that I clipped from *S&S*. During this time, the Allied Forces were steadily turning the "bulge" the other direction.

---

1. CT-12 Unit Journal 0001-2400 27 January 1945. [Record Grp 407; File 304-INF (12). Office of the Adj Genl. National Archives and Records Administration, College Park, MD — referred to hereinafter as 407/304-INF (12)]

2. CT-12 Unit Journal 0001-2400 28 January 1945 (Ibid).

3. Ralph's recollection about the pay increase is correct and equates to about $160 today. The CIB was a prized badge first instituted by then-commanding general of the Army Ground Forces, Lt. General Lesley J. McNair, on October 27, 1943, and was made retroactive for deserving GI's to 1941. In order to receive the CIB, a soldier must have demonstrated "satisfactory performance of duty in action against the enemy" according to 1943 War Department documents. A year later, the requirements further clarified that the recipient had to have been engaged in "ground combat against enemy forces." In other words, simply being present in a war zone did not entitle one to the CIB. "Combat Infantry Badge CIB." *HRC: 'Soldiers First!'*, July 7, 2022, https://www.hrc.army.mil/content/Combat%20Infantryman%20Badge%20CIB

4. Ralph's memory was on-target. 12th Regiment CP was located in the village of Gilsdorf January 21-27, 1945.

It was also during this period that I had an experience similar to Willie's in Mauldin's cartoon (**above**). A barn with hay in it was a sought-after place to spend the night. In many cases, the animal barn was simply part of the house.

I can also vouch for the authenticity of Mauldin's cartoon (**next page, left**) on the subject of women. No conversation of any length failed to include at least some mention of this long-lost but not forgotten part of life!

January 28, 1945, we spent the night at Moinet, Belgium, just outside Bastogne. The next day we were off again on a long trip to Burg Rueland, Belgium, right on the border of Germany.

We left Burg Rueland, Belgium February 1, 1945, crossing into Germany through the Siegfried Line, and stopping for a few days at Elcherath, Germany. The "bulge" was now fully reversed.

The Siegfried Line was indeed a formidable barrier. In addition to the concrete "dragon's teeth" to prevent crossing by mobile units, the Germans had steel-reinforced concrete pillboxes covering all the approaches to the barriers. Many of these pillboxes were disguised as farmhouses or barns with wooden roofs built on top.

On February 5, 1945, we arrived at Schweiler, Germany, where I had the distinguished (or dubious — you choose!) honor of being made Private First Class!

I also became one of the most ardent supporters of sulfa drugs when a medic sprinkled a little of the powder on a 4-inch festering shinbone sore I had received during the German break-through. This was my first encounter with the "wonder drug." The place healed marvelously within a few days.

We continued our push into Germany, moving northeast to Buchet, Germany on February 8, 1945. The *"Keine Warmestube"* card (**above, right**) was tacked to the door of the small house where the Radio Section billeted February 8, 1945, to March 2, 1945. I've no idea now as to its meaning.[5] But it did remind me of Mauldin's cartoon on the next page. It also calls to mind that I did not always feel comfortable in such confiscated quarters.

---

5. It literally translates as, "No warming room." Vermittlung means "placement" or "negotiation." Warmestube originally referred to a heated room, where in the winter needy, especially homeless, persons, could temporarily stay. Perhaps the card was left by the displaced German owner in hopes of preventing invading troops from over-running their home in search of shelter from the cold and snow — if so, it obviously didn't work!

"Careful. Th' toilet seat's booby trapped."

We had just come through ground previously taken by the 4th Infantry Division in September 1944, just before I joined the 4th. The litter strewn across the country-side — dead animals and twisted wreckage — gave mute testimony to the many battles fought through this area — by the 4th when it first took it, by the Germans in their last great counterattack during the Bulge, by the Germans fleeing back into their country and by the pursuing American forces.[6]

At Buchet we came to a halt for nearly three weeks waiting for supplies to be brought up so the drive could be resumed. A few days of decent weather (a warming spell) brought a miserable mess of mud. Our supplies had to be dropped by air daily. It was the first time I had witnessed airdrops. Very impressive.

But the snow and cold weather soon returned. During this wait, Weltzien taught me how to play chess and the game became a favorite pastime with many of us. We played on a small board sent to Welt-

---

6. 12th Regiment historian Gerden Johnson pronounced this zone "the filthiest area the 12th had ever fought through. Melting snow revealed bodies of both German and American soldiers, frozen where they fell. Hundreds of dead cattle littered the fields, along with carcasses of dead horses. Human excreta were deposited in corners of rooms where the fighting had been at such close quarters that leaving the buildings was an invitation to death." (Johnson, p. 309).

zien by his family. It had sunken holes for the pieces to fit down into to keep them from becoming displaced when we had to move suddenly.

We had a nice rest here at Buchet, which helped prepare us for a hard battle over the city of Prum, Germany that loomed ahead.

Above: Map from from *Stars and Stripes* showing our position in January of 1945 (inside circle).

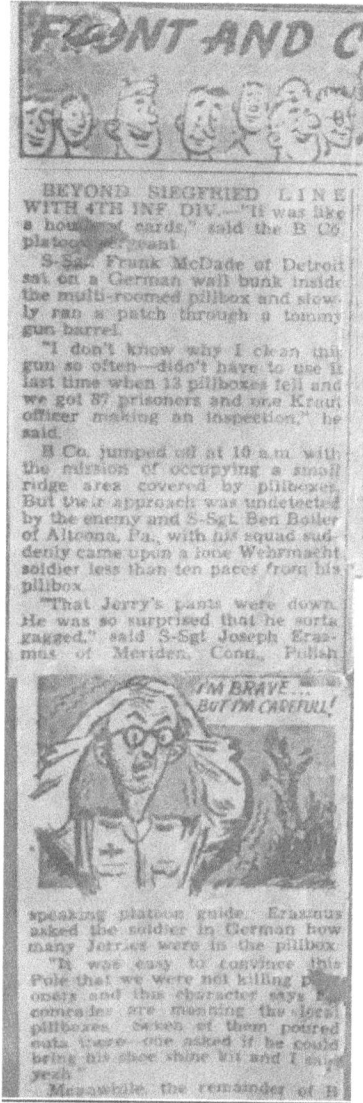

BEER TOKEN FOUND IN CAFE

Our radio section billed in the Café Schmidt-Reuter, where I picked up a beer token as a souvenir (**above**). The regimental command post was housed in another building about a block and half away from us (**below, bottom image**).

Above: I Clipped this "Front and Center" article from *Stars and Stripes* during our movement into Germany in February. War isn't any fun, for sure. But sometimes you just have to laugh at the Devil — it drives the Devil and his buddies crazy and helps the good guys get through it all with a little sanity.

HEMSTAL - Panorama - La Gare

I snapped this photo (**left**) of the Dragon's Teeth that we passed near Brandscheid, Germany on 1 February 1945. The innocent-looking farm buildings in the left background are actually Kraut pillboxes in disguise.

Above: 12th Regiment CP in Burg Rueland, January 28, 1945. We (radio section) billeted in the nearby house on the left. I almost broke my neck tying to string wire to the CP from the structure on the right.

'We're Going to Hang Our Washing...'

reversed.

The Siegfried deed a formidable b addition to the con teeth" to prevent c mobile units, the G steel-reinforced co boxes covering all to the barriers. pillboxes were dis

Above: I cut this out from my copy of *S&S* about our advance into the Fatherland.

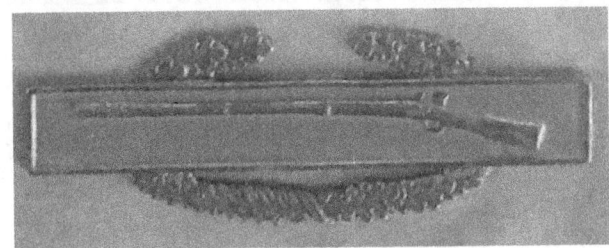

Above: I received my Combat Infantryman's Badge in Gilsdorf while the Regiment was engaged in combat operations in Luxembourg. A few days later, its mission completed, our division was transferred out of this sector and sent north into Belgium to become part of the force that would cross the Siegfried Line and drive on Prum, Germany and the Kyll River.

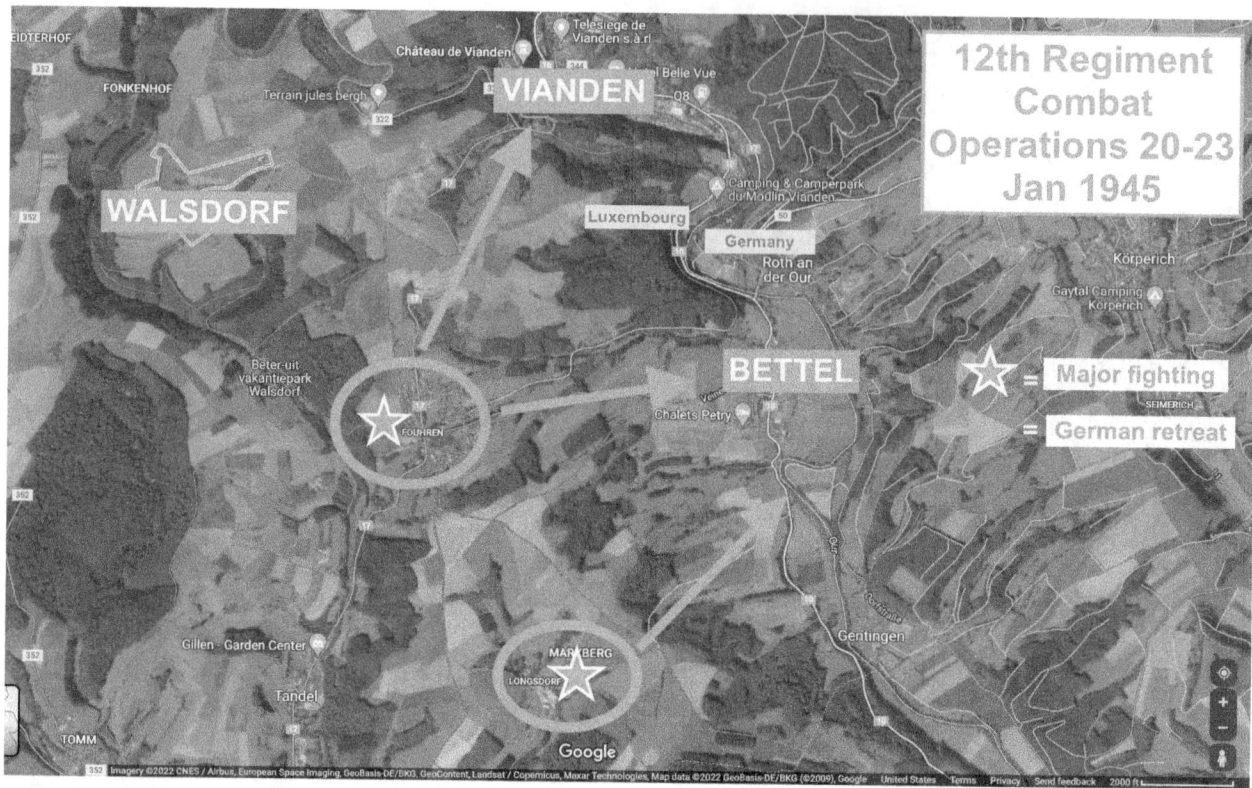

Above: A satellite view with graphics that show the lines of attack by the 12th as they chased the weary, battered German remnants out of Luxembourg. The stars indicate locations of sharp engagements, and the arrows indicate advances by the Warriors as they pursued the retreating Germans toward the Sauer River. These mop-up operations took place while Ralph was attached to regimental command posts at Ermsdorf (January 19-21) and Gilsdorf (January 21-27), about two miles southwest of the fighting shown here.

Above: *"At Buchet... a few days of decent weather (a warming spell) brought a miserable mess of mud. Our supplies had to be dropped by air daily. It was the first time I had witnessed airdrops. Very impressive."*

Above: A pre-war photo of the Café Schmidt-Reuter in Hemstal, where Ralph and the rest of the radio section billeted 28 December – 17 January 1945.

Above: Regimental Command Post in Burg Reuland in photo taken in 2019. The structure that housed the radio section no longer stands — it would have appeared in left foreground.

Above: David stands in front of the 12th Regiment Command Post in Hemstal, Luxembourg as it appears today. The word *"Verainsbau,"* painted over the large door, translates, "Club Building."

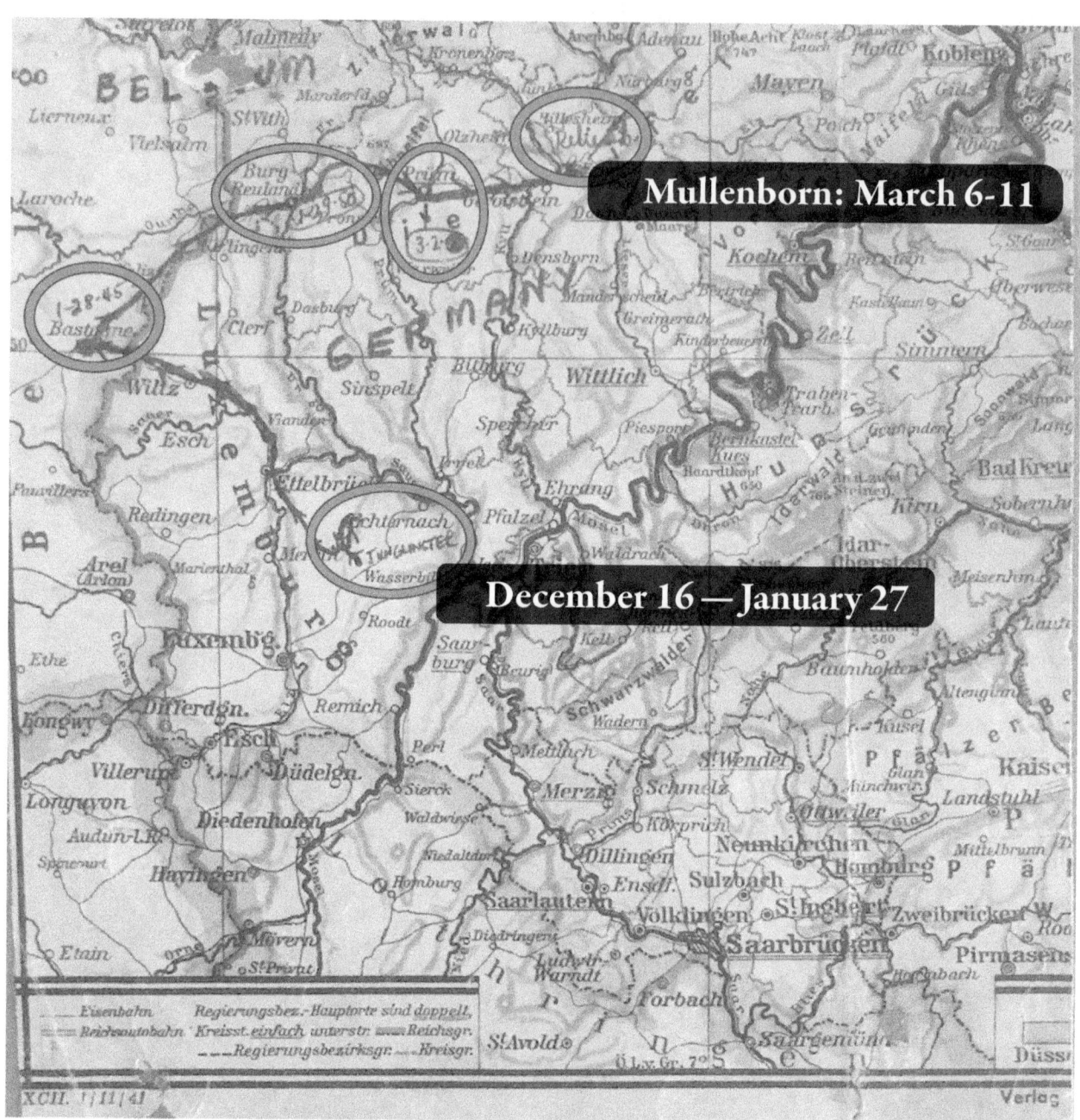

Mullenborn: March 6-11

December 16 — January 27

8. Ankle Deep in Death: Assault on Prum

"... human refuse was ankle deep on all the floors. Krauts littered the rooms — dead..."

Somewhere along the way, I ripped a map (**facing page**) from the geography book of some German student. On it I have marked my travel from mid-December 1944 to early March 1945 — from the time we disengaged from the Bulge until we took Prum. Since I never knew where I was going, I tried at least to know where I'd been!

We were at Bucher,[1] awaiting the assault on Prum, Germany. The assault began March 1, 1945,[2] and the next day we were setting up Regimental Headquarters in that very devastated city. We took over a three-story building (one of the few still standing) which apparently had been used as doctors' offices judging from the paraphernalia scattered about. The Germans had obviously made a last desperate stand in this building. They must have been pinned up in it for some time because human refuse was ankle deep on all the floors. Kraut soldiers littered the rooms (dead, of course) and their gear was everywhere. We cleaned up as best we could, threw all the junk out, and bedded down.

---

1. That is, Buchet. I have discovered similarly mis-labelled town names in no less official documents than US Army After-Action Reports and even in the *12th Infantry Yearbook*, so I give Ralph a pass on this one.

2. It is easy to see how Ralph's memory may have conflated the assault on Prum — which actually began in earnest on February 11 and was largely completed by February 17 — with the taking of Neider Prum ("South Prum"), a village situated immediately adjacent to and southwest of Prum. The 12th had been assigned to the southern flank of the division during the campaign. While the 22nd and 8th Regiments moved on Prum itself, the 12th remained in its southern sector, cleaning out pockets of German resistance in the nearby villages and countryside. On March 1, two battalions of the 12th entered Prum, and on March 2 launched an attack on Neider Prum, capturing the city by the next day; this is no doubt the attack which Ralph references in his account. The 12th IR Command Post moved into Prum, as Ralph recounts, on March 2, to oversee the advance on Neider Prum. With Prum secured, the VIII Corps pushed east across the Prum River in pursuit of the increasingly disorganized German resistance, and then wheeled northeast toward the Kyll River. By March 11, German resistance had virtually ceased, the front became quiet, and the 4th ID was relieved and sent to R & R in France. See map, "Movement of the 12th Infantry Regiment and Regimental Command Posts, 8 February — 5 March 1945." (page 115) For those interested in a deeper dive into the VIII Corps campaign through the Schnee-Eifel of February 1 — March 11, 1945, these are excellent resources: (1) Record Group 407; File 304-INF (12). Office of the Adjutant General. National Archives and Records Administration, College Park, Maryland. (2) "https://imwesten.com/drive-on-pruem/; and (3) History of the Twelfth Infantry Regiment in World War II, by Colonel Gerden F. Johnson, Chapter XIV — "History Repeats Itself."

It was still bitterly cold and snowy during the Prum campaign. Wingert's cartoon (**below**) most accurately expressed our feelings about the weather. The only thing good about the freezing temperatures was the bodies didn't stink (among other things).

The Germans had boobytrapped many of the bodies of their fallen comrades, leaving them for us to turn over and get blown up. The streets and woods were full of bodies. One young German I noticed particularly — seemed only a boy. Looked as though he had never shaven yet. His uniform was brand new. This was probably his first (and last) battle. He lay on his back just as he had fallen from a bullet through the stomach. His eyes were still open (as were just about all the dead) staring glassily up at me. He was so solidly frozen that when I turned his body over with my foot, the body remained in the same position. I ran through his pockets and took out some photos he had. They were (1) a picture of the soldier himself in a bright, pretty uniform; (2) probably his girlfriend; and (3) his mother. The boy's face bore little resemblance to the photograph of the neat soldier as it was now a dirty blue, still contorted with the agony he likely suffered before passing out. I could not help feeling sorry for this kid — surely he did not know for what he had fought so bravely.

Feeling sorry for a dead soldier — German or American — at this time was unusual for we had seen so many. Callous we had become — what else could be expected when soldiers on both sides used the frozen dead to stack around foxholes for protection.

At Prum, the Eleventh Armored Division[3] came through the rubble to take up a sector of the still advancing front. Our regimental headquarters group left Prum March 5, 1945 and spent that night in Wallersheim, Germany. On March 6, we moved up to Mullenborn where we were when advanced elements of the Division reached Adenau, Germany.

Our mission accomplished, the Division was relieved on March 11, 1945, and we journeyed to Bleialf, at which point we entrained on March 12, 1945, for a real rest in France.

---

3. At noon on March 2, the Commanding Generals of the 4th ID and the 11th Armored Division realized that the bridgehead over the Prum River east of the city was too narrow for the advance of the ground troops. They decided to wait until the bridgehead could be enlarged by the engineers of the 4th Division, and the 11th Armored would pass through the infantry in Prum and press the attack on March 4. (http://www.11tharmoreddivision.com/history/march_after_action.htm). See also http://imwesten.com/photos-then/.

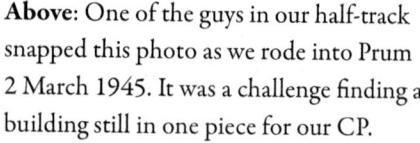

Above: One of the guys in our half-track snapped this photo as we rode into Prum 2 March 1945. It was a challenge finding a building still in one piece for our CP.

Above: I vividly remember moving down this muddy road through Sellerich as we moved on Prum, Germany in March of 1945.

Men are sniper hunting as Prum is slowly cleared out.

PRUM, GERMANY
3-2-45
COMMAND
POST

PRUM, GERMANY
3-2-45

PRUM
3-2-45

Prum, Germany

Soldiers of the 4th Infantry Division in Prum on March 1 — the day before Ralph arrived to set up the Regimental Radio Section billet on March 2 — 5. **Above:** Photos taken from a similar vantage point that graphically show the devastation the assault had brought on the city.

At left is a modern view taken from same street. Notice that other than the church steeple, the buildings are all post-WW2 — the flattened city was almost entirely rebuilt from the ground up after the war.

Prum
17 June 2018

Left: These youthful German prisoners captured at Prum were luckier than the poor lad whose frozen body elicited Ralph's pity and sympathy.

Above: Tanks of the 11th Armored Division maneuver at Fleringen, between Prum and Wallersheim, east of the Prum River on March 4, 1945. Engineers from the 4th ID had to enlarge the bridge that spanned the Prum River so that the advance could continue. Pending completion of that task, the 11th, as Ralph wrote, *"came through the rubble to take up a sector of the still advancing front."*

Left: The Wallersheim command post building as it appears today

Wallersheim Regimental CP
October, 2019

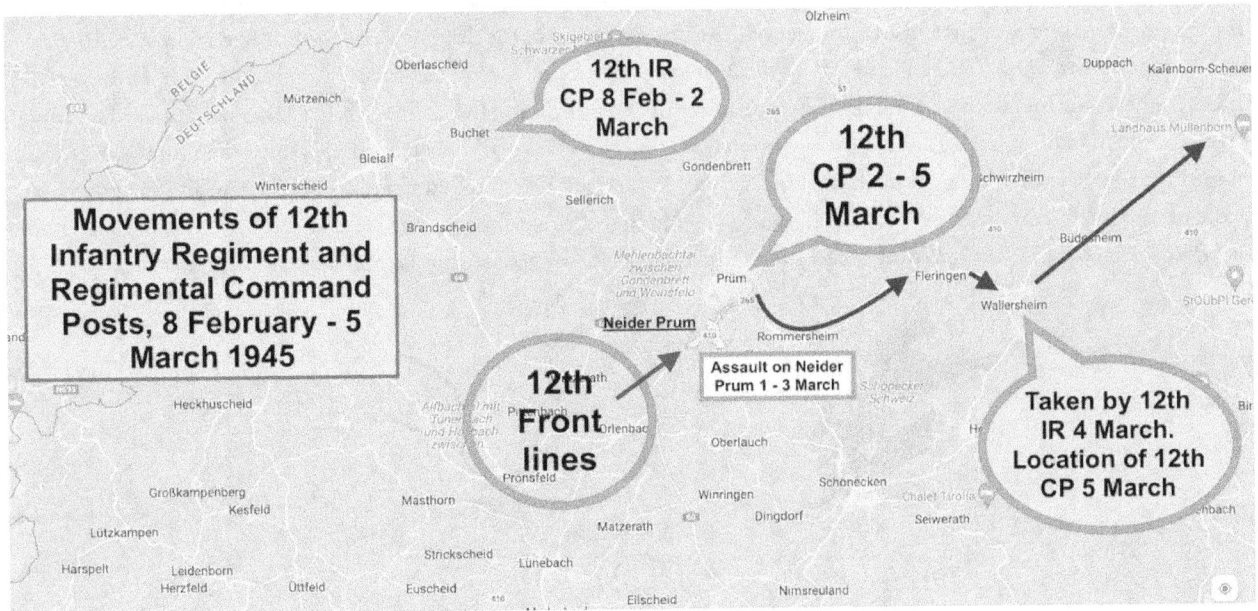

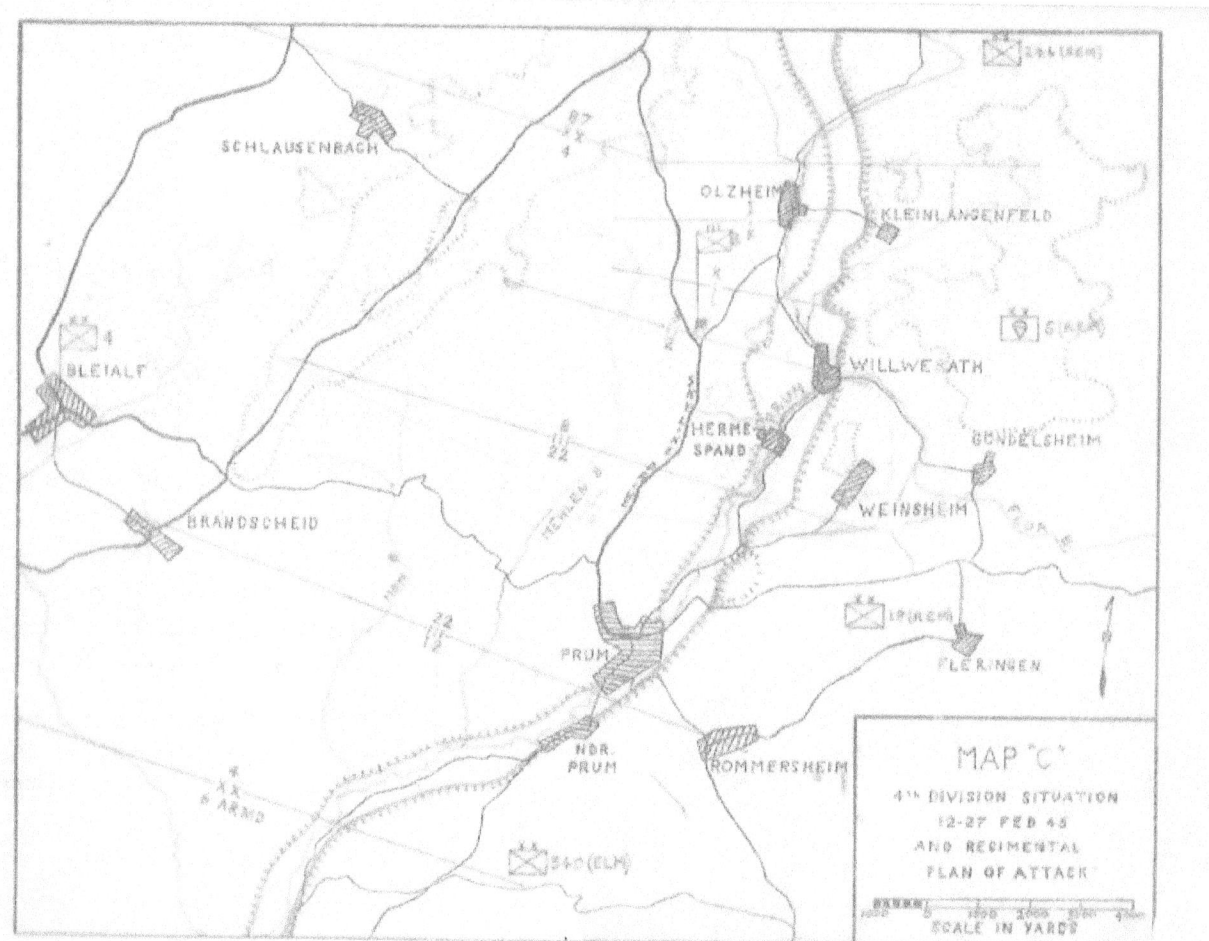

The map **above** shows the assigned attack zones of the three regiments of the 4th Division: the 8th Regiment aimed at Hermespand in the north, the 22nd at Prum in the center, and the 12th (Ralph) at Neider Prum in the south. However, all three fought in and/or occupied Prum proper at one time or another during the campaign. The Division crossed the Prum River in early March, with the 12th advancing through Rommersheim and Wallersheim. The Ivymen then turned northeast toward the Kyll River, reaching Mullenborn and Adenau. At that point, active German resistance ended, and the Division was sent into R&R in France on March 11. **Note:** Most accounts of this campaign refer to it as the "Two Rivers Campaign," or "The Fight for the Rivers," referring to the Prum and Kyll Rivers, key geographical landmarks during the battle.

Source: http://imwesten.com/pruem-river-crossing/

Left: Modern view of the same street into Sellerich that Ralph followed in March 1945.

Above: Modern street view in the village of Mullenborn. Ralph and the rest of the Radio Section of HQ & HQ Company were here when the Division was ordered to assemble at Bleialf, where they entrained for southwestern France and a week of R&R.

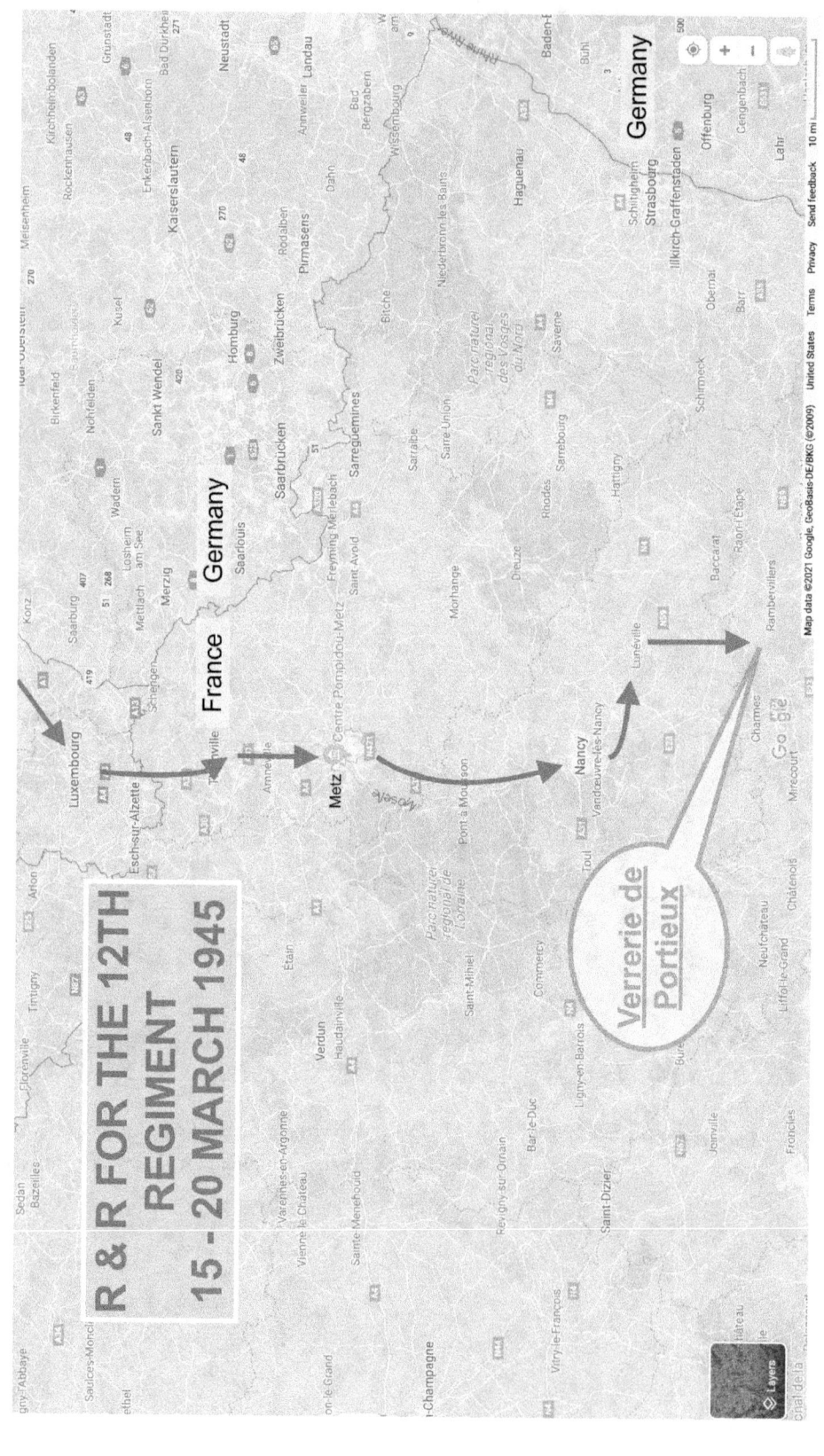

R & R FOR THE 12TH REGIMENT
15 - 20 MARCH 1945

Verrerie de Portieux

9. Hot Showers and Clean Beds

"… we enjoyed a wonderful week's rest, with hot showers, clean beds, hot food, and relaxation."

Following the tough fight around Prum, our Division was relieved on March 11, 1945, and we journeyed to Bleialf at which point we entrained on March 12, 1945, for a real rest in France.[1]

On the map on the next page I have outlined (as I went) our move from the Hurtgen Forest (1) to Luxembourg (2), and from Luxembourg back into Germany through Prum (3), all in white. Then in black can be seen our general route down into the Alsace-Lorraine province of France, via the city of Luxembourg, Metz and Nancy, France. In the vicinity of Luneville, France, our Regiment was assigned to the small village of Verrerie de Portieux (4). From there, after the end of a week of R&R, the black lines show our route north and then east into Germany (5). The 4th, I learned later, was now within the jurisdiction of the Seventh Army.

The trip was a memorable one via famous French "Forty and Eight" (forty men or eight horses) boxcars. The train stopped every few minutes and at each stop the men would jump off, run out to the fields to attend to nature's needs, then run back to the train with handfuls of straw, looking toward more pleasant slumbers that night.

During this trip I sustained my worst mishap of the war. At one of the frequent stops, I got off the car a few minutes to relieve myself. Just then, the train started off again. Several of us were running alongside the cars with arms outstretched to be helped aboard. In my usual nimble-footed way, with my eyes fixed on the car I ran full tilt into the end of a stack of steel rails. The greatest force of the impact was against my left knee. The blow knocked me unconscious for a second or two after which I managed to get to my feet just as the last car came by and I was whisked aboard. The pain was so severe I lost consciousness again and the next time the train stopped, one of the boys went ahead and came back with the medic. The kneecap was laid bare, but the wound was a fairly clean one. It was later I learned the kneecap was cracked.

In addition to the kneecap injury, the fates had conspired to make the trip more miserable by afflicting

---

1. The troops were sent southward in several shifts. Ralph and the rest of the HQ Company, along with the 1st Battalion, left around midnight March 12, and arrived in their assigned rest area in Verrerie at 8:30 am the next morning, March 13. [Memo — Action Against Enemy, Reports After/After Action Reports, 28 April 1945, p. 4. (407/304-INF(12)].

me with a bad case of the "GI's," sometimes known as the "midnight runs," but more technically, dysentery. The bum leg made it a real nightmare.

At Verrerie de Portieux, France, we enjoyed a wonderful week's rest with hot showers, clean clothes, hot food, and relaxation.[2] After our first good shower, the Radio Section put its collective head together to talk things over — probably planning out our "vacation itinerary" during this rare and all-too-brief break from the stress of combat.

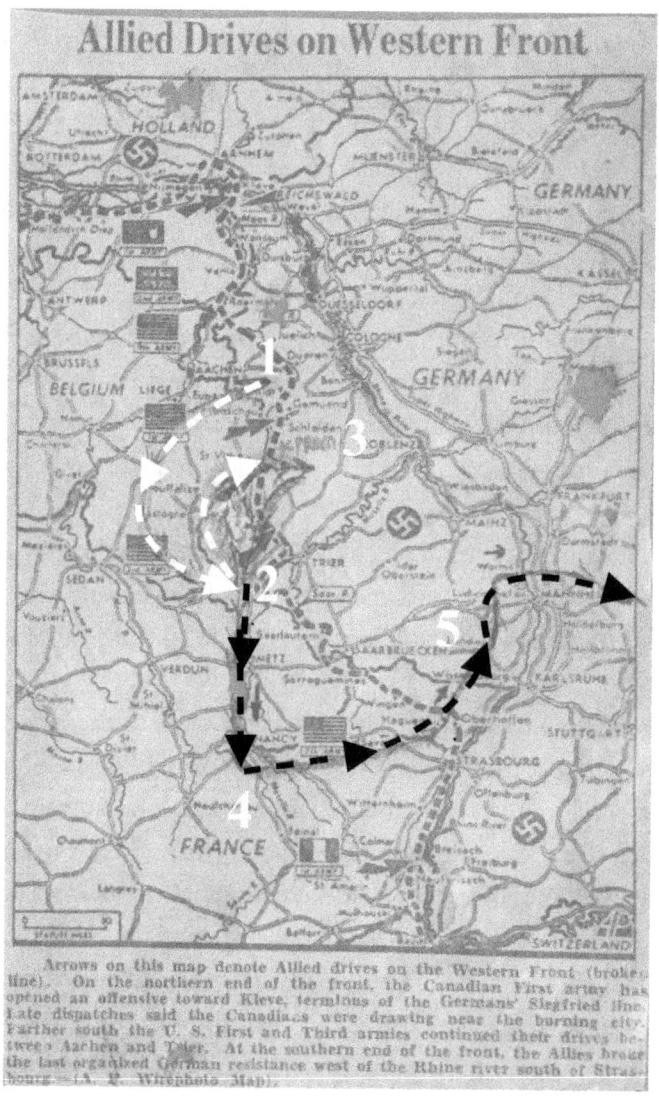

_____

2. The brass took good care of the men who had performed so bravely from the Hurtgen to the Bulge to Prum: "Showers being enjoyed by all troops, as well as clean clothes for all. Regiment CO insisted that … all men are billeted well and with a program of hot food, movies, and other recreations. Spirits are high." [(Unit Report for 14 March 1945 (407/304-INF(12)].

Our next order of business was to get acquainted with the ladies of the area. Here we surround the prettiest one to be found, Rose Marie (last name unknown), and a couple not so pretty — but females!

During my stay, I enjoyed playing "Papa" to three French children, who stood around and eyed the leftovers in our mess kits with pitiful longing. Needless to say, they fared as well as we did while we were there.

We enjoyed this quiet, pastoral little get-away for five glorious days. It would prove to be a rest we would badly need for the final push that lay ahead.

Above, left: Three French kids enjoy the attentions of a quite willing American GI "Papa." These photos were taken March 15, 1945, at Verrerie de Portieux.

Above: Radio Section has a meeting of the minds after our first hot shower in Verrerries. I've forgotten the exact topic of conversation, but it must have been interesting, March 15, 1945, at Verrerie de Portieux.

* * * * *

The Army could not have chosen a better place for its battle-weary 12th Infantry Regiment to rest, re-cover, and re-charge than Verrerie de Portieux. Not only were the men exhausted physically and emotionally from unremitting and intense combat since entering the Hurtgen Forest early in November of 1944, but their equipment, vehicles, armaments, and materiel were in desperate need of repair, refitting, and replacement. Here, in a tiny, bucolic village far away from the thunder and threat of combat, the men could fully relax, (and train some so as not to lose their edge) and generally disengage mentally from the stressful rigors of war.

Verrerie de Portieux had been liberated by the 14th Armored Division on November 7, 1944 — the same day that the 12th entered the nightmarish Hurtgen. Verrerie had been home since the late 17th century to a manufactory of exquisite crystal pieces (*Verrerie* is French for "glass"). The Cristallerie de Portieux operated from 1690 until 1996, when the company was merged with and relocated to the Cristallerie de Vallerysthal in the village of Troisfontaines, some forty or fifty miles northeast of Verrerie.

Above: Diorama at the National Museum of the United States Air Force in Dayton, Ohio, of USAF POW's being transported to a stalag in a French "40 and 8" boxcar. First used in WWI, the boxcars were also used in the Second World War to transport soldiers, prisoners, and other personnel by both sides in France. (https://www.nationalmuseum.af.mil/Visit/Museum-Exhibits/Fact-Sheets/Display/Article/196342/french-forty-and-eight-railroad-car/)

Above: Cold, dirty, and war-weary 12th Infantry dogfaces board "forty and eight" boxcars March 12, 1945, in Bleialf, bound for a week of hard-earned R&R in Verrerie de Portieux, a trip that Ralph called for a number of reasons, none of them good, "memorable."

Left: Turn of the century postcard of the train station in Verrerie de Portieux

Right: It was a deeply emotional moment to sit on the same steps where Ralph and his comrades had posed with the local ladies so many decades earlier.

Left & Below: David poses in the same spot in Verrerie de Portieux where Ralph *"played papa to three French children."*

David Miles
Verrerie de Portieux
October 19, 2019

Ralph

Above: "After our first good shower, the Radio Section put its collective head together. I've forgotten the topic, but it must have been interesting." Below: "Next order of business was to get acquainted with the ladies of the area…"

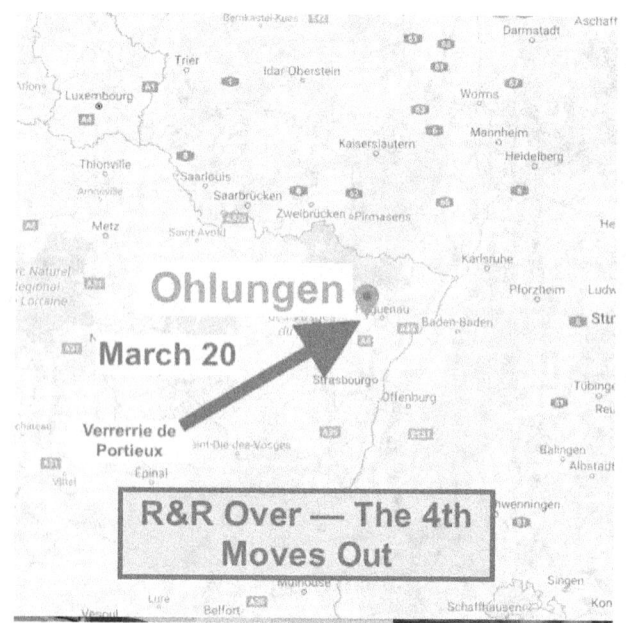

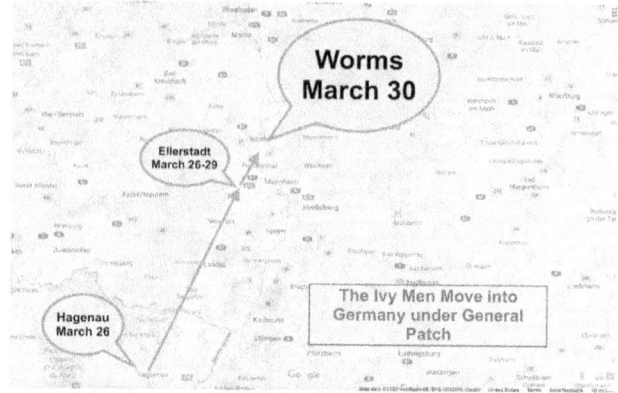

Left: Fourth Division left its encampment at Verreries de Portieux in France on March 20, 1945, and moved through Olungen and Hagenau to Ellerstadt, Germany. From there the Ivies advanced to Worms (**above**), where the Division, now under General Patch, crossed the Rhine River in Germany. The Ivymen then drove eastward toward Wurzburg (**below**).

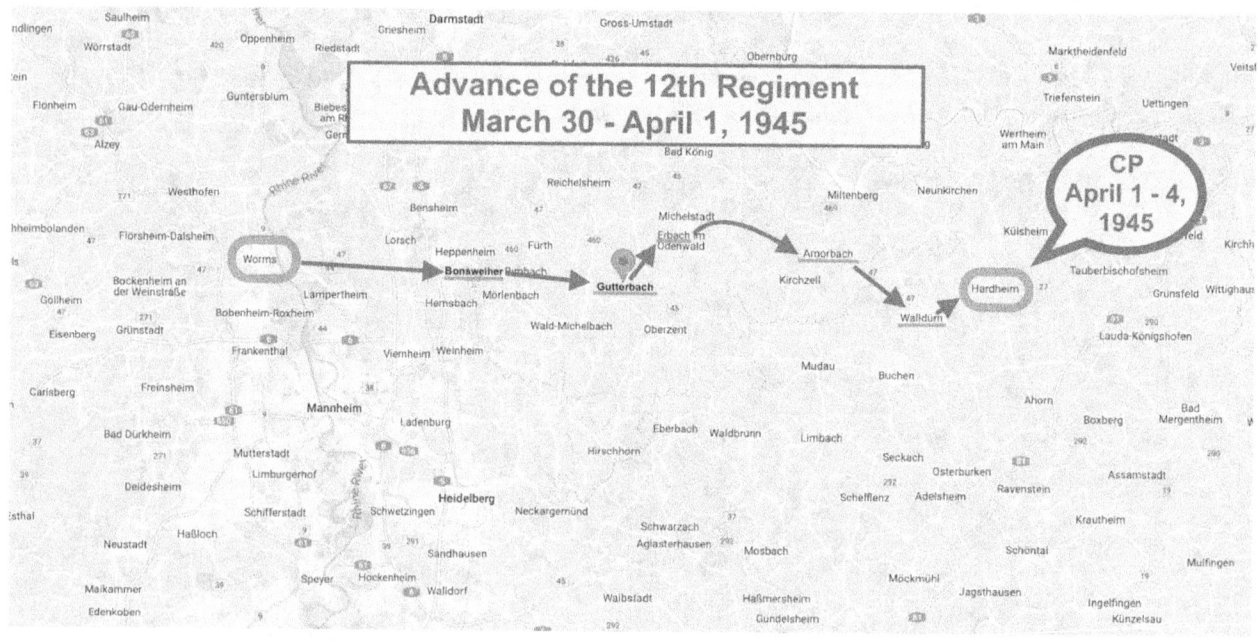

10. Entering the Belly of the Beast

"We were reversing the 'bulge' made by the Germans in our lines during the Battle of the Bulge…"

No Parkway, This

These beautiful parkways Commissioner Bob Moses built in New York and Long Island don't permit trucks. This is one of Hitler's own superhighways, but the trucks belong to the 9th Armored Div., which are highballing east. —Signal Corps Photo

Reluctantly we left our rest area on March 20, 1945, and traveled unopposed eastward toward the Rhine River, stopping at Ohlungen, France for the night. This is in the province of Alsace-Lorraine, near Hagenau, about ten miles from the Rhine River. We were moving into the lines with the Seventh Army (under General Patch).[1]

The people of this area spoke German almost exclusively, probably due to the fact that it was once German territory (I think), which was lost as best I recall to France at the close of World War I.

On March 26, 1945, we passed through Hagenau. From this point, we turned north and crossed the German border through the Siegfried Line to Ellerstadt, Germany. We were still unopposed, and Regimental Headquarters remained at Ellerstadt three days. Then we continued northward, driving on Hitler's superhighways (**left**), the first I had seen and a truly remarkable type of roadway.

As we went zipping through the country, stopping occasionally in a hay-filled barn to grab a little sleep, I felt somewhat like the character in Wingert's cartoon (**next page**)! Seemed as if we could never get enough sleep or rest to get past the "doze" stage.

On March 30, 1945, our outfit turned east again and crossed the Rhine River at Worms,[2] Germany. I could not help but be reminded of the "Diet of Worms," the meeting in 1521 at which Martin Luther was condemned as a heretic.

---

1. Alexander M. Patch had served with distinction in the Pacific Theatre in 1942, personally leading US troops in the capture of Guadalcanal. Eisenhower transferred Patch to the European Theatre of Operations, where he led the invasion of southern France in August of 1944 as commanding general of the Seventh Army, succeeding Mark Clark in that position. Patch died of pneumonia in November of 1945 (https://en.wikipedia.org/wiki/Alexander_Patch).

2. Thereby joining in the follow-up to Operation Undertone — also known as the Saar-Palatinate Offensive — that had been tasked with

After a brief layover at Bonsweiher,[3] we were again in heated contact with retreating Germans. We moved swiftly in pursuit, moving in one day, April 1, 1945, a Sunday, through Gutterbach, Erbach, Amorbach, and Waldurn, winding up at Hardheim[4] at the end of that day. We could hardly process the prisoners we took fast enough — disarm them and send them to the rear. We were continuing to reverse the "bulge" made by the Germans in the "Battle of the Bulge."

Leaving Hardheim April 3, we were off again, with very favorable weather. The next day, out of Gaubuttelbrunn, our Regiment took the lead. Freed slave laborers thronged the roads and mutely searched for food. Pale, gaunt, solemn, these people were among those most deserving of a long rest and nourishment. They pillaged the towns and villages and overran gardens and German homes, but we did nothing to stop

driving the Germans across the Rhine and clearing the Palatinate region of western Germany of enemy resistance 15-20 March 1945. The Seventh Army was composed of a number of US divisions, and coordinated as well with the French First Army to the south under General Jean de Lattre de Tassigny that was advancing against the Germans from east of the Rhine (https://dbpedia.org/page/Operation_Undertone).

3. As in the case of Bucher/Buchet, all of the official Army records refer to a brief stop in "Bonsweiler," during this drive, where the 12th Regiment CP was set up for a short time. Ralph's handwritten inscription on the picture of the CP building reflects the same erroneous spelling. The actual name of the village is "Bonsweiher," with an "h." The US Army can defeat the Axis juggernaut, but they struggle mightily with spelling!

4. Unit reports from April 1, 1945, record that German jet-propelled planes strafed and bombed the Regiment at Hardheim. Had the Germans developed this technology sooner — and if they had had sufficient jet fuel to power the planes — the outcome of the war might have been quite different! [Unit Report for 1 April 1945, dated 2 April 1945, from Record Group 407 (File 304-INF12).]

them unless they resorted to actual violence against civilians. We could well understand their feelings toward their erstwhile captors and torturers.

We were meeting only occasional opposition now as our drive into central Germany gained momentum. Moving forward every day, sometimes five miles, others fifty miles.

By April 13, 1945, through Bernsfelden and Reidenheim, we reached the outskirts of the city of Wirzburg, spending that night at Bieberehren,[5] about five miles from Wirzburg. Here we heard of the death of President Roosevelt. Much talk among the troops concerned whether or not Truman could fill Roosevelt's shoes and what effect this tragic event would have on the outcome of the war. It seemed pretty obvious to us the Germans were about done.

"When they run we try to ketch 'em. When we ketch 'em we try to make 'em run."

At Freundenbach, Germany, on April 14, 1945, I received a cablegram from Eleanor's sister Fleeta telling me of Ralph Junior's birth a week earlier on April 6.

Our direction now turned south, and our advance picked up steam. We continued taking prisoners by the score — so many, in fact, we could do little more than take away their rifles and equipment and send them to the rear.

---

5. Captured by the 2nd Battalion on April 11-12. The Regiment continued driving southeastward, brushing aside sporadic German resistance in villages along the route Ralph describes (Johnson, pp. 325ff).

Births

Pfc. and Mrs. Ralph J. Miles announce the birth of a son on April 6. He has been named Ralph J., Jr. Mrs. Miles was formerly Miss Eleanor Hobbs. Pfc. Miles is somewhere in Germany.

Above: Ralph Junior's birth announcement from the Savannah *Morning News*. All Eleanor knew was that I was "somewhere in Germany" — this was, of course, before the days of cell phones, Face Time and Skype. The uncertainty was really tough on those at home.

On April 16, we drove about five miles to Adelshofen; on April 17, approximately five kilometers to Schweinsdorf; on April 18, four kilometers to Kirnberg.

April 19, we went through the picturesque walled city of Rothenburg[6] and then on to Schillingsfurst. At Schillingsfurst, we took over a beautiful castle belonging to a German prince.[7] It was a castle in the feudal style, having a drawbridge and everything. In my letter to Eleanor the next morning, I described it:

"It is really beautiful, sitting high above all the surrounding area. As I write, I'm sitting on a bench in the sunshine. Spread out before me is a vast and beautiful panorama. The bluff falls sharply away almost at my feet and you can see for miles. Trails of dust mark the passage of vehicles and smoke spiraling toward the sky marks the residue of battle..."

Marvelous paintings bedecked the walls in the castle, along with wonderfully carved swords. While here, I "liberated" a chess set, wooden.

Alas, I was fated to spend only one short night in this quaint and mythic setting before the call came to resume the fight against our retreating but still dangerous foe.

---

6. Rothenburg was captured without a shot due to the intervention of Assistant Secretary of War John J. McCoy, who realized the historic importance and beauty of the medieval city. Per McCoy's instructions, six (or five, depending on the source) dogfaces from the 12th IR "volunteered" to walk into the city under a white flag and demand that it be surrendered or face destruction; if the GIs did not return safely by 6:00 pm, the reduction of the city would commence. The small German garrison chose wisdom as the better part of valor and surrendered the city to the Red Warriors, whose lead elements entered the city April 17. Ralph and the HQ & HQ company passed through the town two days later, as Ralph wrote. (Johnson, pp 329-338; also https://military-history.fandom.com/wiki/Military_history_of-rothenburg_ob-der-tauber)

7. In fact, according to Johnson, the Warriors captured the Good Prince Hohenlohe himself! (Johnson, p.338)

Above, left: "Swede" Carlson and me. **Above, right:** Me holding a "contraband of war" — a tasty chicken, maybe? Both photos were taken during our drive toward Wirzburg in April of 1945.

Above left: Radio half-track squad sits for a "portrait" during a rare pause in our drive through Germany in the spring of 1945. Yours truly is squatting on the right, next to pal Weltzein. **Above right:** Mauldin's very accurate take on why Krauts liked to surrender to souvenir-hungry GI's. I got mine direct from their erst-while owners who no longer needed them!

Above left: The 12th Regiment CP in the village of Bernsfelden, as it appeared April 7, 1945. **Above right**: Present-day view of Bonsweiher regimental CP building. The structure housed a school in 1945. Today it is a museum — unfortunately, it was closed when we visited the site.(Both the official Army records and Ralph misspelled Bonsweiher as "Bonsweiler.")

Above is a modern-day view of Amorbach, one of the villages Ralph and his fellow GI's raced through in the early spring of 1945. The timber-and-wattle, Hansel-and-Gretel appearance of this and so many other German villages is like something out of a vintage Disney cartoon movie.

Above Left: Nancy Miles poses in the plaza at Hardheim in October of 2019. Ralph's unit arrived here at the end of a day of rapid advance on April 1, 1945. He and his comrades remained in Hardheim until April 3, at which time the division continued its advance southward against aggressive but poorly coordinated German resistance.

Above right: David stands at the outskirts of Gaubuttelbrunn, where on April 4, 1945, Ralph and his fellow troops encountered a sight that provoked both their pity and their anger: *Freed slave laborers thronged the roads and mutely searched for food. Pale, gaunt, solemn, they pillaged the towns and villages and overran gardens and German homes, but we did nothing to stop them.*

* * * * *

Encounter in Bernsfelden

We pulled our rented Opel into the narrow, deserted village streets of Bernsfelden on a cool, grey fall afternoon. I parked the car and Nancy immediately got out and disappeared around some low buildings in search of the object of our visit: the building — if it still stood — that had been used as the 12th Regiment command post in early April 1945.

I remained behind, having noticed an older lady sweeping her porch across the street from where we had parked. I decided to take a chance with my (very) broken German to ask her if she knew where the American soldiers had been housed in their village during the war.

Bernsfelden Regimental CP
October 7, 2019

Herr Uhl

Nancy Miles

The frau listened quizzically for a moment and then motioned for me to wait. She turned and disappeared into her house. Moments later, a man who proved to be her husband emerged from the inside, pulling on his jacket. I managed to communicate to the patiently attentive *der Herr* my name and the purpose of our visit. He introduced himself as Herr Uhl, and he indicated not only that he knew the building that I was looking for, but that he would be willing to take me to see it.

We walked together down the street and circled the same block of buildings that Nancy had rounded several minutes earlier. Sure enough, we found Nancy already standing in the parking lot of the modest and still neat and well-kept *schulhaus* that had housed the 12th Regiment Command Post some seventy-five years before. Herr Uhl, Nancy, and I chatted about my father's movements with the American Army through the area during the war, and I told him of our sentimental journey following in his footsteps. Remarkably, between his patience and my "German," we managed not only to communicate, but to bond as friends.

After taking several pictures of the schoolhouse and of ourselves with our new-found friend, we thanked Herr Uhl for his time and helpfulness and turned to leave. However, Herr Uhl, instead of saying goodbye, motioned toward an adjacent cemetery. It was clear that he had something else that he very much wanted to show us, so we followed him the short distance to the quiet village graveyard.

Bernsfelden Cemetery

We made our way through the tombstones and stopped in front of a monument erected to honor German war dead from both world wars. Herr Uhl pointed past the monument and across an open field stretching from the village.

"My father was killed there during the battle in 1945," I interpreted from his words and his gestures.

He then pointed to two names engraved on the plaque on the memorial: Nicholaus Uhl Senior — his grandfather, killed in WW1 — and Nicholaus Uhl Junior — his father, listed as *"vermisst"* ("missing") in WW2. Nicholaus Junior was later determined to be KIA.

As I stood there, reflecting somberly with Herr Uhl, it struck me that my own father or one of his comrades may have been the very ones who killed Herr Uhl's father. Yet even with that mutual realization, neither of us felt any anger or resentment in the moment. I didn't feel any hate for his father, nor did I find it strange or awkward to honor the memory and sacrifice of Nicholaus Uhl for his country. In the very same way, Herr Uhl betrayed in neither his words nor his demeanor that he had any problem conversing warmly with the son of the very one who may have slain his own father in battle.

It was a profound moment — both moving and enlightening. Courage, self-sacrifice, willingness to fight — and die — for what one believes and in defense of one's country, deserves to be remembered with gratitude and respect, regardless of the ideology of the leaders who started the conflict. No matter who the Chancellor or President is, or what the politics are, the names on that plaque represent husbands, fathers, sons, and brothers who stood between their families and mortal danger. That should mean something. In fact, it should mean a great deal.

And we must always remember them with honor and gratitude.

Here's to the memory of a brave and patriotic man, Nicholaus Uhl, Junior, and to his kind and hospitable son, our friend, Herr Uhl.

Above: The broad fields that lay outside the village cemetery in Bernsfelden. After showing us the names of his father and grandfather engraved on the War Memorial in the graveyard, Herr Uhl pointed to these fields, where his *Wehrmacht* father was killed by the advancing American army — the 12th Regiment ... my father? — in April of 1945.

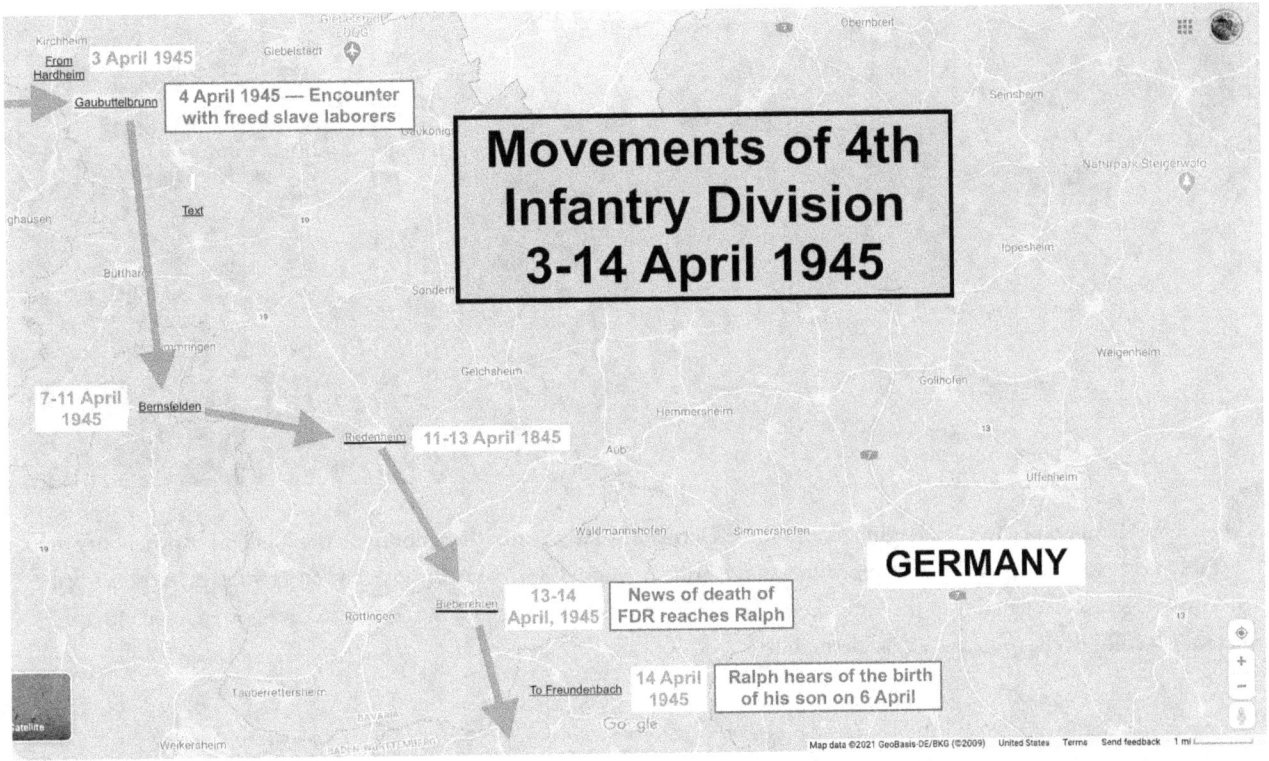

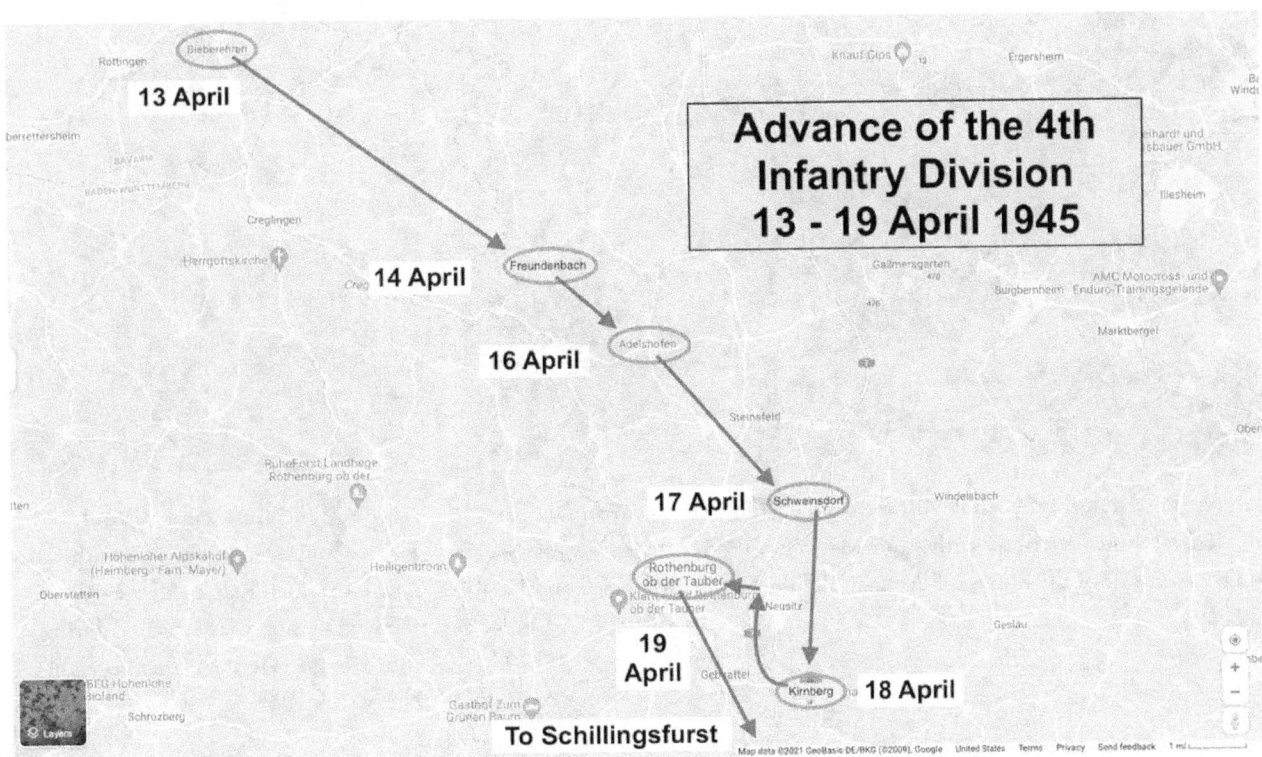

Prisoners coming into Rothenburg's colorful town square
April 17, 1945.

Above left: Members of the 12th Regiment and some German POW's enter Rothenburg ob der Tauber ("Rothenburg on the Tauber") town square with German prisoners on April 17, 1945. Ralph came through the city two days later on April 19. **Above right**: Nancy stands in the same town plaza October 7, 2019. The city is thought to have gotten its name from the reddish color of the roofs, as *roth* is German for "red."

Above: Lead units of the 4th Division cross the Rhine River at Worms as part of General Sandy Patch's Seventh Army March 29, 1945. After a three-day stay on the west side of the Rhine at Ellerstadt, Ralph's Company crossed on March 30.

Above: General Alexander "Sandy" Patch, Commander of the Seventh Army, to which Ralph and the rest of the Ivy Division were attached during the drive into Germany that began in late March 1945.

The Prince, the Castle and the 12th Infantry

Ralph spent only one night at Schillingsfurst Castle — April 19, 1945. But the experience clearly made a deep impression on the flat lander from rural south Georgia. Ralph's rhapsodic description in his letter to Eleanor of the bucolic beauty of its setting, its striking architecture, and the ornate *accoutrement* adorning its interior walls reveal his awestruck wonder at the palace.

And little wonder, for Castle Schillingsfurst is a spectacular specimen of the ancient *schlosser* that are strung like so many jewels along the famous *Romantische Strasse* ("Romantic Road") that bisects southwestern Germany from Wirzburg to the Austrian border. The imposing mountain-top fortress perched above the eponymous village nestled below has been the home of the politically powerful Hohenlohe family from as early as the fourteenth century. The property remains in the family and is the residence of the current Prince, Constantin, 12th Prince of Hohenlohe-Schillingsfurst, who lives with his family in the south and west wings of the castle (the vacated north wing is open to the public for tours).[8]

When Nancy and I arrived at Schillingsfurst Castle during our 2019 journey following Ralph's footsteps, we were as awe-struck by the baroque beauty of the place and its sylvan setting as Ralph was when he set up shop there nearly eight decades earlier.

Our docent, Maximilian (seen above with Nancy), or "Maxi" as he preferred to be called, was a pleasant, soft-spoken young man who regaled us with all manner of fascinating historical information about this medieval manor. During his presentation, Maxi recounted how the Prince and his family had abandoned the north wing of the castle on April 19, 1945, never to return, and left the rooms and furnishings just as we see them today. The entire family moved across the courtyard enclosed by the U-shaped structure to occupy the south and west wings.

8. https://www.wikiwand.com/en/Hohenlohe. https://second.wiki/wiki/schloss_schillingsfc3bcrst

However, neither Maxi nor any of the other guides knew why the Prince and his family had left the north wing so suddenly on that spring day long before. But I knew. I told them of the occupation of the castle by the 12th Regiment of the 4th Infantry Division, and how the regimental command headquarters — including Ralph and his fellow GI's — was set up in those very rooms on April 19, 1945, dispossessing the unfortunate Prince and his brood! I also told them of Ralph's rapturous description in his letter to Mom of what he saw inside the castle and of the gorgeous views from the castle heights overlooking the village. They were quite thrilled to add this new information to their guide sheets, and they asked that I send them a copy of Ralph's letter to include in their records, which I was only too happy to do.

As we strolled through the *schloss* marveling at its beauty — and imagining in our mind's eye the incongruous picture of battle-begrimed Ivymen tramping across the delicately inlaid floors and plopping themselves in exquisitely cushioned antique chairs — we encountered the lady pictured **above.** Hailing from the nearby town of Neuenstein about thirty miles west of Schillingsfurst, she spoke English beautifully (which many Europeans, at least those in urban areas, do, just as Ralph discovered while in Luxembourg City). Nevertheless, she did have enough of an accent that I could never quite understand her name. As I was too embarrassed to ask her to repeat it over and over, I spell it here phonetically as close as I could make it out: Hedda (or perhaps Helga) Denninger, with a "hard" *g*.

Hedda, overhearing me tell our hosts of Ralph's arrival at the castle with the 4th Infantry Division in 1945, chimed in with, as the radio commentator Paul Harvey used to say, "the rest of the story."

It appears that, as the 12th Infantry Regiment approached the village and the castle, they set up to besiege and if need be, reduce the structure. Some of the local boys, in a show of foolish youthful bravado, decided to resist the "Ami's" by firing their rifles at the Yanks from the castle roof. The servants of the family were horrified, fearing that the boys' actions might provoke a destructive bombardment of their beloved home. So, the servants dashed up to the roof where the boys were, wrested their weapons away from them and frantically waved as many white sheets at their would-be attackers as they could find. The Americans, seeing the white banners, held their fire and took their objective without firing a shot.[9]

I filmed a video of Hedda in which I complimented her on her delightful personality, wonderful English and fascinating tale.

She responded with, "And I am glad to meet the son of the officer who entered the castle and did not kill it!"

Above: William "Bill" Chapman, Captain of Company E, 2nd Battalion, 12th Infantry Regiment and commander of Task Force Chapman that captured Schloss Schillingsfurst. By war's end, Chapman had risen to become the S-3 at 2nd Battalion Headquarters.

---

9. The taking of Schillingsfurst by the 12th Regiment was spearheaded by Captain Bill Chapman of the 2nd Battalion, Company E. The Company was formed into Task Force Chapman as the unit passed through Wohnbach, directly north of the schloss, and Bellerhausen, a village situated northwest of the castle and its princely occupants. As the Ivymen approached the town and castle, they took artillery fire from the direction of Bellerhausen. In response, the Task Force mounted its tanks and tank destroyers April 19 and sped toward the castle, under Chapman's command. They motored up the mountain on an unpaved trail at the base of the north wing, captured Prince Hohenlohe and commandeered the prince's quarters in the north wing of the castle for the Regimental CP. As Gerden Johnson observed, for Task Force Chapman, the whole operation "was ... a field day" (p 338.). No doubt the ease with which the GI's took over the edifice was helped along by the sheet-waving house servants mentioned by Frau Denninger. For those interested in more details of this and other operations conducted by Captain Chapman and Company E, both at Schillingsfurst and elsewhere, I recommend *Battle Hardened: An Infantry Officer's Harrowing Journey from D-Day to VE Day*, authored by Bill's son, Craig S. Chapman.

Above: Nancy Miles taking photographs inside the courtyard of Castle Schillingsfurst. **Below**: Nancy at the palace entrance described by a marveling Ralph in his letter to Eleanor: *"It was a castle in the feudal style, with a drawbridge and everything."*

Left: *"As I write, I'm sitting on a bench in the sunshine. Spread out before me is a vast and beautiful panorama. The bluff falls sharply away, almost at my feet and you can see for miles."*

Below: David stands at the overlook that Ralph admiringly wrote Eleanor about.

Marvelous paintings bedecked the walls of the castle, along with wonderfully carved swords.

Below: Wooden chess set that Ralph "liberated" from Prince Hohenlohe at Schillingsfurst.

Above left: David stands before the entrance to the north wing of Schillingsfurst Castle — the door through which Ralph and his comrades would have entered the building on April 19, 1945. **Above right**: Nancy stands at the front of the entrance to the south wing, to which the Prince and his family moved April 19, 1945, and where they still reside today.

Left: The startling sight that greeted Ralph as he walked in the north wing of the castle to help set up the CP! This stuffed bear had been felled by one of the Prince's daughters years before the war. It has stood menacingly in this spot ever since.
Right: The road likely used by Task Force Chapman to ascend to and capture Schillingsfurst Castle April 19, 1945.

Above: Likely direction followed by Task Force Chapman to sieze Schillingsfurst Castle April 19, 1945.

Above: As Ralph entered the north wing and turned to the right, he ascended these stairs and no doubt marveled at the intricate carvings in the stucco walls depicting various Roman emperors.

Left and above: Photos of the interior rooms of the north wing — looking just as they did on April 19, 1945, when Ralph and the other 12th Regiment HQ personnel set up their operations. I don't know where the 8th and 22nd Regiments had their CP's, but I can only imagine their envy of the regal surroundings of their companion outfit!

Below: Busts of the children of the Prince who were living in the Castle when Ralph and the HQ company occupied the north wing.

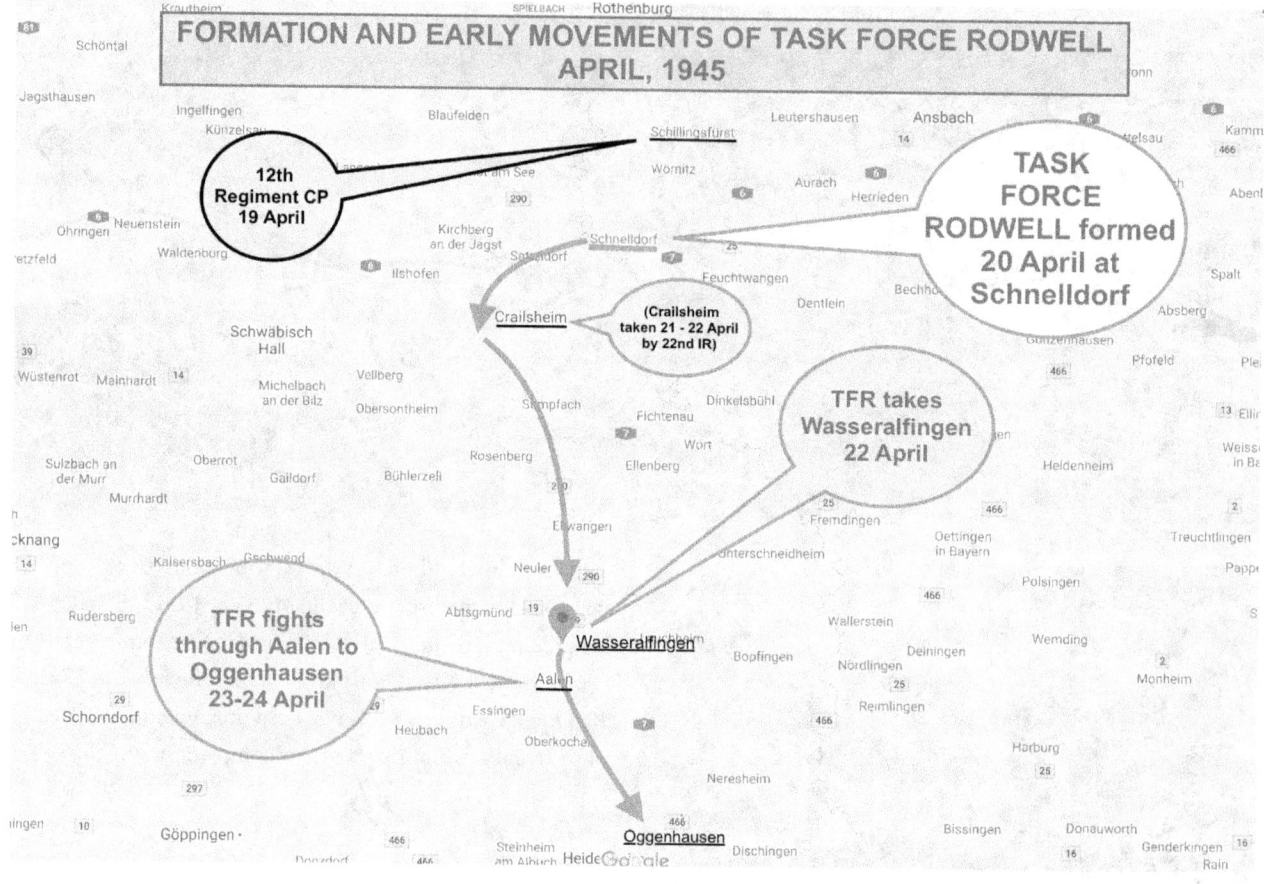

FORMATION AND EARLY MOVEMENTS OF TASK FORCE RODWELL APRIL, 1945

12th Regiment CP 19 April

TASK FORCE RODWELL formed 20 April at Schnelldorf

(Crailsheim taken 21 - 22 April by 22nd IR)

TFR takes Wasseralfingen 22 April

TFR fights through Aalen to Oggenhausen 23-24 April

11. Task Force Rodwell

"... I didn't sleep twice in the same place, and, in fact slept very little at all."

April 20th, I left the rest of the company and journeyed to Schnelldorf (about fifteen kilometers) to join "Task Force Rodwell."[1] This task force consisted of motorizing a battalion of the 12th Infantry Regiment and support elements to use as a spearhead ahead of the main body of the 4th Infantry Division as it continued its advance southward through Bavaria. The Task Force was commanded by Assistant Division Commander General James Rodwell. I was selected as radio operator in the Colonel's half-track (Colonel R.H. Chance). We were not to rejoin our complete outfit until we reached the Danube. I didn't sleep twice in the same place, and, in fact, slept very little at all.

The Task Force fought through Crailsheim[2] and on southward, aiming at the fair-sized city of Aalen. The record I was trying to keep of my travels is very sparse during this period to the end of the war because we moved so often and so fast.

One place I recall fairly vividly — Wasseralfingen, a small town just north of Aalen. We arrived in

---

1. This was Rodwell's second task force to lead, his first having performed so well during the advance on Prum in February and March of 1944 that Rodwell was awarded the Distinguished Service Cross for his efforts. A task force, as Ralph indicates, consisted of selected units from various parts of the Division, all of which are motorized, including infantry. The mission for TFR was to move in advance of the main body of the Army, blazing a pathway of destruction and weakening the opposing forces. TFR was composed almost exclusively of Red Warriors, plus various tank, and other motorized battalions from the Division, including the 522nd Field Artillery Battalion. It must have been a signal honor to be selected as Colonel Chance's personal radio operator — Ralph had obviously proved his mettle in combat conditions not only to the satisfaction of his fellow squad members, but to that of his commanders as well. R.H. Chance had taken over command of the 12th from Colonel Luckett during its terrible days in the Hurtgen. Regrettably, Ralph had little time to record more of his interactions with Colonel Chance — but he was a busy man indeed as TFR moved relentlessly and rapidly thorough the heart of Hitler's Reich over the next five days.

2. This is one of the few times that Ralph's ordinarily sharp memory and keen mind for detail was a bit off. TFR encountered small arms and rocket fire, as well as other obstacles, at Crailsheim, so they simply by-passed the town and sped southward, ensuring that the demoralized Germans stayed on their heels and had no chance to regroup and resist the Allied onslaught. Meanwhile, Crailsheim itself was taken after some hard house-to-house fighting by the 22nd Regiment April 21-22. (Johnson, p. 340) [After Action Against Enemy, Reports/After Action Reports, dated 9 May 1945 (407/304-INF (12)].

Wasseralfingen right on the heels of the departing Krauts and even before the townspeople had taken in The Nazi streamers and flags. Among my bulkier souvenirs, several of these may be found.[3]

Shortly after we entered Wasseralfingen, I happened to run into a German in about the finest uniform I had seen thus far. I said to myself, "Surely he must be a Field Marshal!" Turned out, he was a fireman! I made him give me the sleeve patch off his uniform and it can be seen at the end of this chapter.

There was a schoolhouse in Wasseralfingen where we spent the night, which was worthy of notice. I quote from my letter to Eleanor of April 23, 1945:

What part of last night I slept, it was in a fairly large schoolhouse, quite a substantial building and handsomely furnished. It had evidently been used recently as a barracks for German soldiers because all the desks were piled in the hallways and double-decker bunks installed in the rooms. It gives one quite an insight into the German mind to look at the murals on the walls ... these walls were decorated with pictures of brutal-looking German soldiers in various poses. There are two which are particularly striking. One depicts a soldier saying farewell to his wife. He holds her in one arm while his other hand firmly clutches a rifle. He is gazing off into the distance with feet apart in the manner of one who thinks to conquer the world. The other is that of a German standing over a kneeling man with one hand clutching the hair of the victim while the other is a closed fist posed high above for a mighty blow. On the German's face is a murderous, contemptuous scowl. In the background are neat, orderly farms and houses characteristic of this land. Perhaps these two murals reflected Nazi education: 'Leave home, wife, possessions. Be brutal, ruthless. But orderly, obedient.'"

April 24, 1945, we passed through Aalen and proceeded to Oggenhausen. The next day, we drove about forty miles to reach Gundelfingen, which was about four kilometers from the Danube River, in the heart of Bavaria. Here I finally completed and mailed to Eleanor and my mother two home-made Mother's Day cards which I had been working on for nearly two months. I was told later that the card to my mother was the last communication she received from me, except my letter of May 2, 1945,[4] before she fell into a final coma.

---

3. See Chapter 15, "War Relics Gallery"

4. See Chapter 12, "Alles Ist Kaput"

Above: Shoulder patch I liberated from a Wasseralfingen fireman April 23, 1945. His fine livery rivalled that of any Nazi Field Marshall I ever saw.

Above: My home-made Mother's Day cards. Top one was for my own mother, Janie. The card beneath I addressed to Eleanor as if Sharron and Ralph Junior were writing the message.

Above: David at WW2 Memorial to German dead in Wasseralfingen, Germany, October 2019

Left: German fireman's uniform bearing the sleeve patch like that confiscated by Ralph in Wasseralfingen.

Left: Brigadier General James Rodwell, Assistant Division Commander of the 4th Division and leader of Taskforce Rodwell

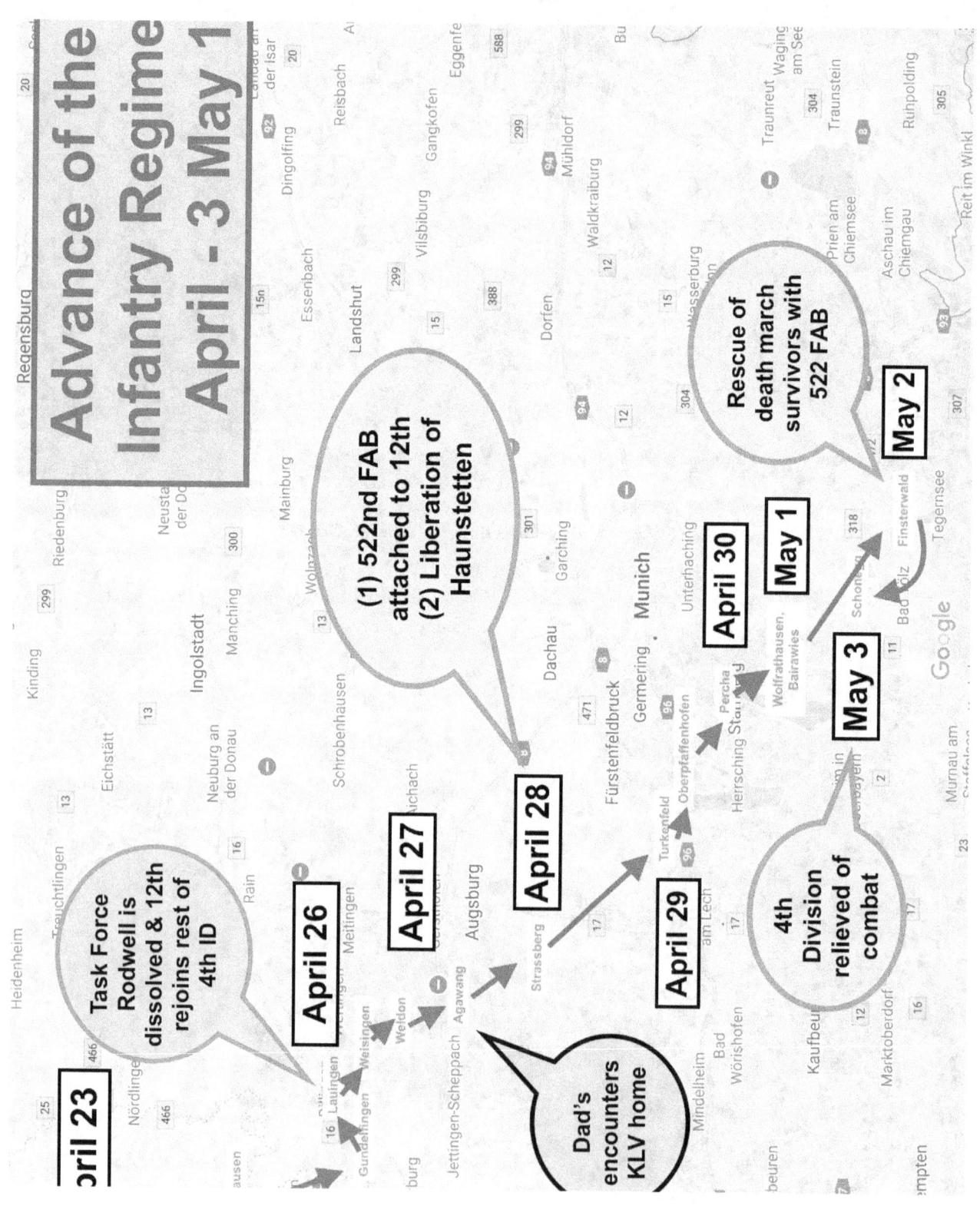

12. *'Alles ist Kaputt'*

"... we freed a whole slave labor camp. We had our hands full just taking over the German soldiers who came out of the towns and woods to surrender, shouting, 'Alles ist kaputt!'... "

On April 26th, at Lauingen, Germany, we crossed the "Blue Danube" — only it wasn't very blue. It was like any other ordinary river to me except that the approaches to it were beautifully landscaped and the riverbanks were bricked in! Also at this city, our Task Force was dissolved, and we rejoined the rest of the Division, our mission accomplished. We stopped that night for a few hours at Weisingen, then, traveling like mad, poured on through Welden, Agawang, Strassberg, Windl, Turkenfeld, Ober Pfaffenhofen, Percha, Bairawies, Finsterwald and finally to Schonegg.

During this southerly drive in the direction of Hitler's Berchtesgaden, we fought through the southwestern outskirts of Munich [(spelled "Munchen" on the German map I tore from a schoolbook (see **next chapter**)] a large and devastated city.

Another day during this trip, we pulled into a large home and found it to be a place where German girls resided during pregnancy under Hitler's "free love" system whereby German girls volunteered to go to the German army camps, sleep with the soldiers, and turn their babies over to the German state.[1] The place was overrun with kids from one to ten years old. We ran the whole lot out and took over. The next morning, before daylight, you can imagine our sleepy-eyed consternation when a whole flock of the kids came tramping through our room shouting in unison, "Heil Hitler!" At first, we thought the Heinies had us for sure!

We were now in the Bavarian Alps. The largest town I can find on a map is Bad Tolz. Across the Alps was Innsbruck, Austria. It was at this point, at Schonegg, on May 3, 1945, that our Regiment was relieved and began to make preparations for a new direction. Rumors were rife that the Germans were preparing to surrender. It seemed abundantly clear to us that the fighting would not last much longer.

Interestingly, during the fighting in the Bavarian Alps, our outfit captured the German General von

---

1. Such homes were called *lebensborn* ("fount of life") facilities by the Nazis. However, the fact is, Ralph and the guys actually encountered a *Kinderlandverschickung*, or KLV, facility, not a lebensborn home. See the end of this chapter under the heading, "We pulled into a large home... overrun with kids aged 1-10..." for a fuller account of this encounter, which occurred at the village of Agawang on April 27.

Rundstedt, who had led the almost successful German counter-offensive known later as the "Battle of the Bulge."[2]

Also, in this area we freed a whole slave labor camp. We had our hands full just taking over the German soldiers who came out of the towns and woods to surrender, shouting, *"Alles ist kaputt"* (*"all is finished"*).

While at Finsterwald, I wrote this letter to my mother on May 2. I had been deeply worried when I left for Europe that my mother, already extremely sick with ovarian cancer, would not survive until I could get back home. Sadly, my concern proved well-founded, as she passed away twenty days after I sent her this letter, the last communication she received from me before she lapsed into a final coma.

[Finsterwald,] Germany, May 2, 1945

Dearest Sweet Mother,

Just got your letter of April 19, Mom. Although I know you shouldn't try to write in your condition, it was so good to see your familiar handwriting again. Of course I knew why you hadn't written before and understand even better than you might think. I'm glad you write your exact condition for I couldn't stand being deceived.

Mary [Ralph's younger sister] says you kinda like my little son [Ralph, Junior] and I'm happy, Mom dearest. Now you are the grandmother of 13 kids! Goodness knows I was proud when the news came. Could hardly keep my joy to myself. And to think he was born almost the exact time and day I was! Really must have brought you back into the dim past of 24 years ago. I know I haven't been the son to you that I could have and should have, but the tragic thing is that no matter how very much we'd like to, we can't change the past one bit. We can only look to the future and ask God for understanding and, receiving it, act accordingly. Someday, soon I sincerely hope, we'll be back together again, enjoying as never before the fresh poignant smell of the Georgia pines standing so tall and stately around the old homestead on the sandy hills. Once again, we'll revel in the joy of baiting a hook with the wriggling worms for an old turtle to gulp down. Perhaps we might even catch one! We'll walk slowly through the woods in the thick pine needle carpet, fighting off mosquitoes. We'll enjoy being rich, if not in money, in the ability to appreciate what God has given so freely to all who will take time to partake!

---

2. Official accounts credit the 36th Infantry Division with the capture of Von Rundstedt at Bad Tolz on May 1, 1945. The 36th was operating in the same area as the 4th at the time. One source even identifies the specific 36th ID soldier to whom the Field Marshal surrendered: 2nd Lt Joseph E. Burke (http://www.texasmilitaryforcesmuseum.org/36division/archives/seigfri/prize.htm). It is possible that Ralph received some garbled information as the word passed through the ranks, especially since the two divisions were operating so close to each other. It is also possible that some elements of the 4th were involved in some way in von Rundstedt's arrest but did not play the lead role in the incident and so were omitted from the official accounts of his capture.

Know it sounds unbelievable but it's snowing outside — has been for several days. This morning there were 2 or 3 inches of snow on the ground. The branches of the trees bowed under their burden as in prayer. Never thought I'd see the day when it would be snowing in May!

Mom, you can get another check from the Office of Dependency Benefits in Newark, New Jersey for the one that got burned. Eleanor has their exact address. Do you know the number of your allotment? Starts with an X. Write these people and tell them what happened and request that they issue you a duplicate one. If you have the # be sure to mention it and if not, tell them you don't know it or they'll write back and ask for it. Get Mary to write it or Eleanor if you don't feel up to it. Do this now, Mom, and I'm certain they will send it to you. If you don't, and they don't receive a copy of the cashed-in check after a certain time, they will write and ask about it themselves. So get a letter off right away for it may take some time.

Tell Mary to bring the little newcomer over to see you often. He can't help but love his grandmother.

I'm well as can be, Mother. Don't worry about anything. Just take care of your precious self. Must close now but will think of you always. God bless you, my dear sweet Mother.

Your loving son,
Ralph

PS: Hope you get the Mother's Day card I made for you by May 13.[3]

R

Insofar as my outfit was concerned, as it turned out, the war was over. On May 4, 1945, we began a peaceful half-track journey northward through Munich, past a place which was to become infamous — Dachau — and into central southern Germany through Nuremberg.

On May 5th, we arrived at a place where we were to spend some time, Sulzbach-Rosenberg, and where we would receive some very, very big news, indeed.

---

3. Janie did indeed get the cards that Ralph hand-drew and inscribed with a Mother's Day message (Chapter 12).

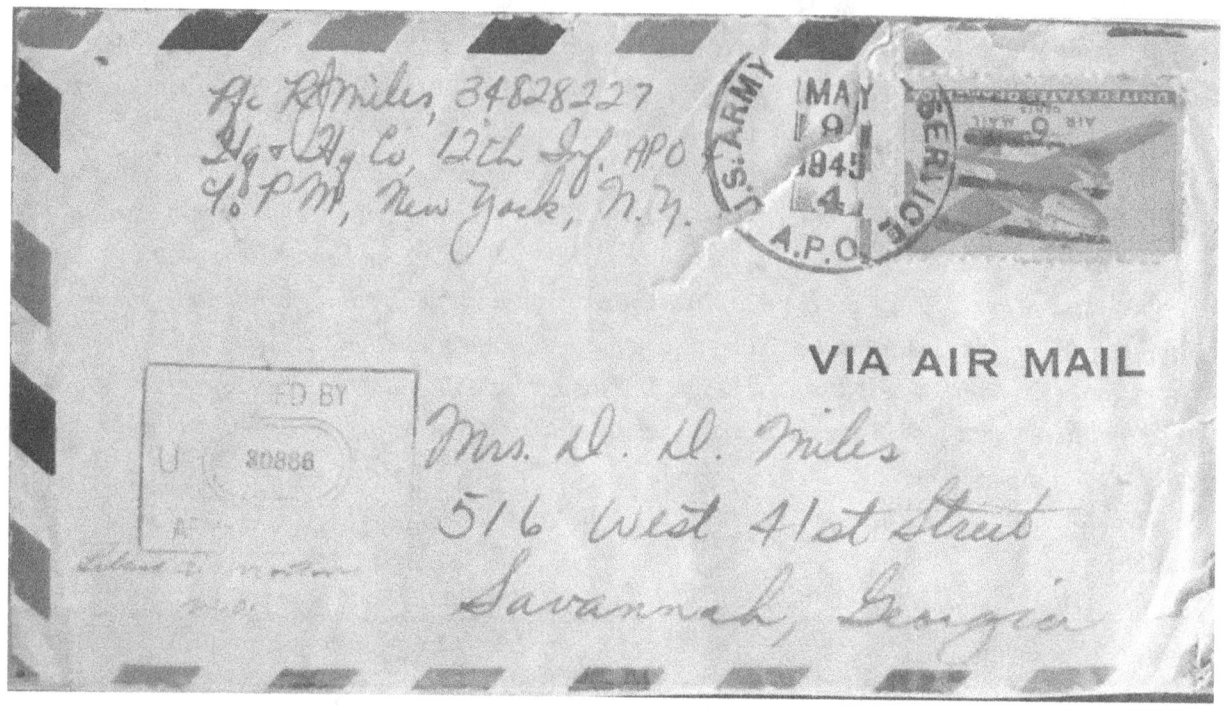

Above: Envelope in which I sent Mother the last communication she received from me before passing away of cancer. It was written while I was near Finsterwald, 2 May 1945.

Left: How the large cities had been devastated by our bombings and shellings! How thankful we were that such destruction had not come upon American cities. These photos were taken, as best I recall, at Nuremberg, during one of our two passes through that once-proud ideological epicenter of the Nazi regime — now completely laid waste.

The blue Danube at Lauingen, April 25.

Left and below: Photos taken the same day Task Force Rodwell crossed the not-so-blue Danube into Lauingen to rejoin the rest of our division. I'm somewhere in the convoy below moving through the city.

I was in one of these trucks

12th Infantry Columns Moving Through Lauingen, Germany, April 25, 1945 to Cross the Beautiful Blue Danube at that Place. Enemy Opposition Was Beginning to Crumble.

Historical photo of Dachau taken same day that Ralph's outfit passed by — May 4, 1945

Above: We passed by Dachau May 4, 1945 en route to Sulzbach-Rosenberg after being relieved from combat in Schonegg May 3.

Above: On a much more pleasant note, here are some post card pictures I picked up along the way showing the Bavarian Alps scenery — all pretty impressive to a South Georgia country boy like me!

* * * * *

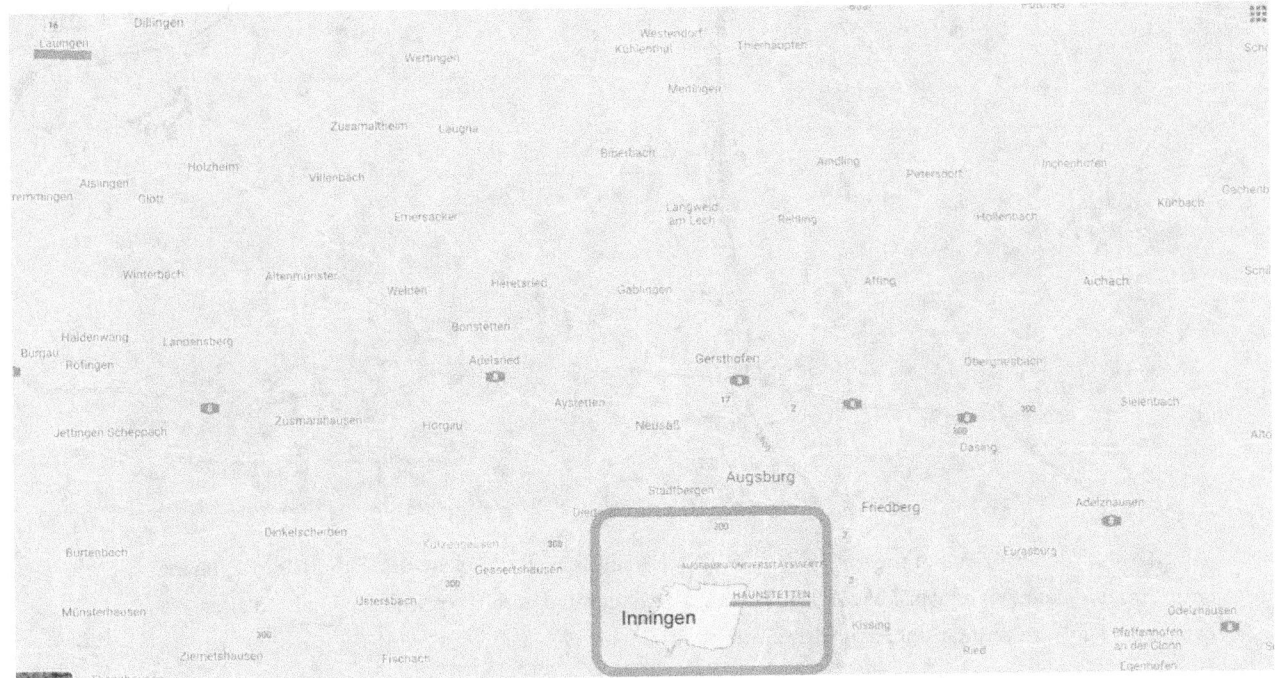

As Ralph's unit passed through Strassberg, a few miles south of Augsburg, the Ivy Division liberated a slave labor camp called Haunstetten, located between Augsburg and the adjoining community of Inningen. A large Messerschmidt factory had been built at Haunstetten, and a large portion of its labor force came from a nearby Dachau satellite slave labor camp.

The factory itself had been bombed by the allies in 1943 and put out of commission. The Holocaust Museum officially credits the Fourth Infantry Division with liberating this camp when it passed through this area April 27-28, 1945.[4]

The Division — and the 12th — had other encounters with Nazi slave camps and camp survivors during its advance, though this is the only one for which the 4th has received "official" credit.

---

4. https://encyclopedia.ushmm.org/content/en/article/the-4th-infantry-division

Lauingen, Germany
October 9, 2019

Lauingen, Germany
October 9, 2019

Above: Modern views of the route taken by Task Force Rodwell and elements of the 12th Regiment through Lauingen on April 25, 1945. The Task Force reunited with the rest of the Division after crossing the river and resumed its drive into Bavaria.

Above: Crossing the "Blue Danube" at the same place where Ralph and the rest of Task Force Rodwell crossed nearly eight decades earlier — and it doesn't appear any more "blue" today than it did to Ralph in 1945.

"In this area we freed a whole slave labor camp."

The question regarding exactly which "slave labor camp" Ralph references in his written account is an intriguing one. Ralph also refers to "Polish slave laborers" on his index cards that he used when speaking to school groups. One possible answer to this mystery is the Haunstetten slave labor camp discussed earlier. However, this camp was outside Augsburg some eighty miles north — nowhere near "this area" — and it was liberated several days earlier, on April 28.

A far more likely answer is that the Red Warriors participated in the liberation in early May of at least two of the many large groups of "displaced persons"[5] scattered throughout southern Bavaria in the chaotic final days of the Third Reich. In late April of 1945, SS camp guards had begun emptying Dachau and its satellite subcamps of their prisoners as the Allies advanced and German defeat became increasingly clear. The inmates were transferred from the satellite facilities to the main camp and then sent out on horrific death marches toward the Austrian border. As American Army Divisions closed in, the Nazis began abandoning their brutalized charges to their fate and fled, often in civilian clothing but still armed and taking advantage of every opportunity to inflict as much damage and casualties on their foes as they could before being killed or captured.[6]

The first of these occurred on a snowy May 2, between Reichersbeurn and Waakirchen, on what is now Highway B-472, when the 3rd Battalion of the 12th Infantry and an attached unit, the 522nd Field Artillery Battalion,[7] encountered a ragged group of 3,000 such survivors who had been left to die by their SS tormentors. The combat lines of the two battalions stretched along the road from Greiling through Reichersbeurn, and thence to Waakirchen. On May 3, the 12th Infantry was ordered out of the area to Schonegg, while the 522nd FAB remained with these poor wretches to protect them from the roving bands of armed SS dead-enders who were still active in the area. Because they were the first to engage with the survivors and remained with them when the Fourth had to move out, the 522nd has received sole credit in official historical records for the rescue. Today, a somber memorial marks the location where the Americans discovered the tattered, starving ex-captives. In 2025, another monument was added at the site specifically to honor the 522nd FAB for its key role in the liberation of the camp survivors.[8]

5. An expression used in Army radio communications referring to the many abandoned groups of slave labor prisoners that the Regiment encountered as it advanced [CT-12 Unit Journal 0001 4 May 2400 4 May [407/304-INF (12)]

6. https://en.wikipedia.org/wiki/Death_marches_during_the_Holocaust

7. https://goforbroke.org/central-europe-campaign. The 522nd was a Nisei (second-generation Japanese American) unit that served with great distinction in the ETO.

8. (a) https://www.smithsonianmag.com/smart-news/remembering-nazi-death-marches-180977751/. (b) https://encyclopedia. ushmm.org/content/en/article/death-marches. (c) https://en.wikipedia.org/ wiki/Death_ marches_during_the_Holocaust. (d) https:// encyclopedia.densho.org/522nd_Field_Artillery_Battalion/ (e) https://www.nvlchawaii.org/522-liberates-dachau-prisoners. (f) Special indebtedness for research on the role played by the 522nd and the 12th Regiment is owed to Beth Reiman, Isami Yoshihara, Leonard

On that same date, elements of the 1st Battalion rescued another group of 300 Dachau camp survivors in the village of Durnbach, located about one mile northeast of Finsterwald, site of the 12th Regiment Command Post, including Ralph and the rest of the Regimental Radio Section.[9] It makes perfect sense to conclude that Ralph had gotten wind of either — or more likely both — of these two horrifying encounters with victims of Hitler's Dachau death factories. While Ralph did not record these events in his diary notes, they clearly made a deep impression on him — so much so that he included them in his speech notes and in his War Album narrative.

On a final note, even in the unlikely event that none of the doughs from the 3rd Battalion directly participated with the 522nd in the rescue of the survivors, the fact that the FAB was attached at the time to the 12th Regiment would no doubt qualify them as "our guys" in Ralph's mind.

Below is a map showing the where the 522nd FAB and the 3rd Infantry Battalion encountered the 3,000 death march survivors between Reichersbeuern and Waakirchen, as well as where the 1st Battalion rescued the 300 ex-internees at Durnbach.

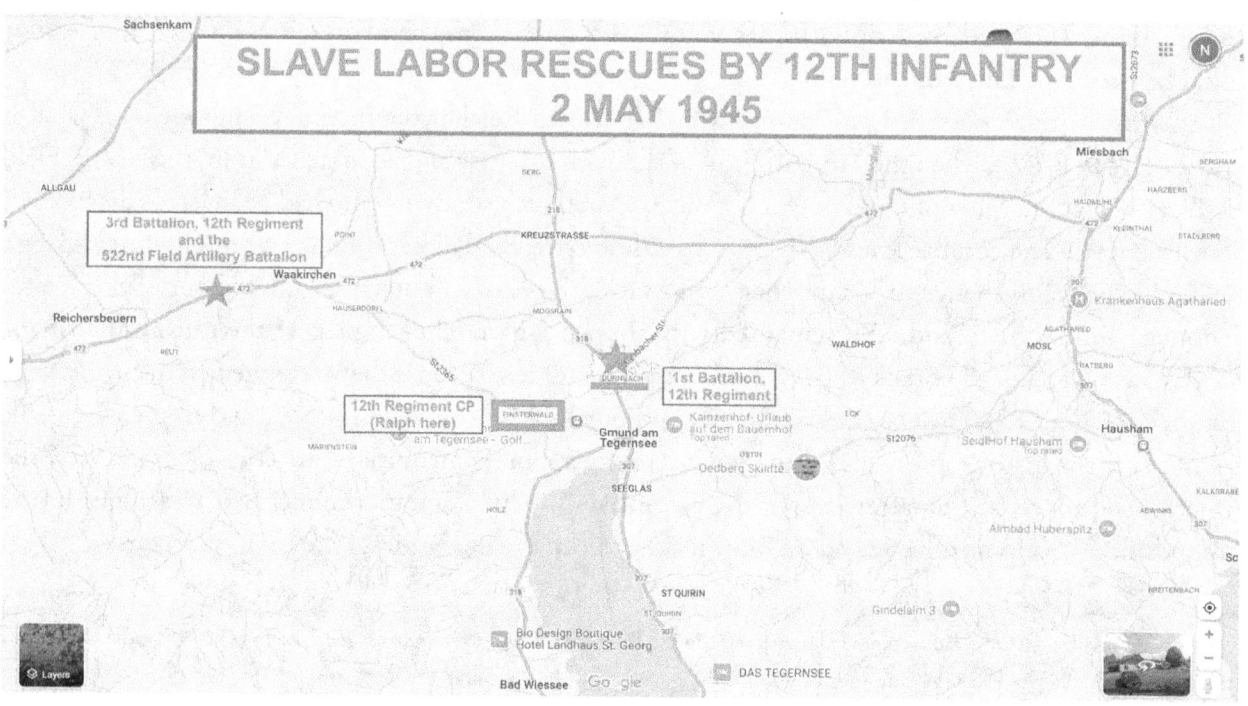

Cizewski and Govan Yee, whose research can be accessed at http://www.ibiblio.org/45wwiiresources/522/522.html?fbclid=IwAR2IlE G6ZaMm5zQJZWBDgdc9OO7DTt-FYbLGaTkQTSxBujoEvzF564e8gt8. (g) https://holocaustjournal.haifa.ac.il/images/Dapim25/ scwartz_english.pdf

9. Schwartz, Eliezer, The Death Marchers from the Dachau Camps to the Alps during the Final Days of World War II in Europe, page 146.

Rescued Death March survivors and one of their saviors, a soldier of the 522nd F.A.B., near Waakirchen

Above: A few of the 3,000 death march survivors from the Kaufering IV Dachau subcamp who were discovered by the Nisei 522nd Field Artillery Battalion and the 3rd Battalion of the 12th Regiment.

Above: Photo of Kaufering IV subcamp at Landsberg am Lech, from which the inmates rescued by the 522nd FAB and the 12th IR were evacuated by SS guards April 27, 1945.

Left: Nancy and David Miles stand next to the memorial at the rescue site of 3,000 death march survivors between Waakirchen and Reichersbeuern. The inscription on the memorial reads, in German:

"Here, in the last days of the war, the ordeal of thousands of downtrodden people from the Dachau concentration camp ended. Many died before liberation."

Above: To the right of the group (featured left) stands the monument erected in 2025 (enlarged at right) honoring those who first encountered the ex-inmates, the 522nd FAB, then attached to the 12th Infantry regiment.

Above: June 2025 — Ralph's descendants stand at the site of the rescue of 3,000 death march survivors near Waakirchen, Bavaria. Behind them is the memorial to the survivors, bearing a plaque with this inscription: "Here, in the last days of the war, the ordeal of thousands of downtrodden people from the Dachau concentration camp ended. Many died before liberation." L-R, front row: Piper Miles, Samuel Jacob Miles. L-R middle row: Eleanor Kate Miles, Adelyn Miles, Ansley Miles, Jessica and Patrick Miles, Madison Hendricks, Colin Hendricks, Allan Miles Hendricks, Ashley Miles Hendricks. L-R, back row: Nancy and David Miles, Emily and Nathan Miles.

"We pulled into a large home ... overrun with kids aged 1-10 ... "

Ralph recounted in his narrative that he and his squad encountered a *lebensborn* home where pure Aryan soldiers had children by carefully selected "pure" Aryan girls. However, extensive research has turned up no Lebensborn facilities in any of the villages in which Ralph's unit spent the night during their drive into Bavaria.

Had Ralph made a grievous error in his otherwise dependable account? Not quite. Internet research on the issue uncovered an article entitled, "Evacuations of children in Germany during World War II" in connection with the village of Agawang, where Ralph spent the night of April 27.[10] The German authorities, much like their counterparts in England during the mass bombings of that country in 1940, ordered the removal of children from the cities that were subject to Allied bombing raids; the children were sent to live with ideologically reliable families in rural areas.

Over time, due to the increasing numbers of children needing housing, the Nazis established large group homes segregated by age: 1-10 years old and 11-14 years old. Bavaria was an area in which a number of these group homes were located. The homes were called *Kinderlandverschickung* ("children to the countryside"), or KLV Camps for short. These KLV camps were not fun places for swimming and recreation. They were grim indoctrination centers run by older *Hitler Jugend* where the children underwent military-style training and immersion in Nazi propaganda.[11]

So, it is clear that Ralph's description of what he encountered does not match a *lebensborn* facility, which would have been filled with infants, young women, and nurse attendants. It *does*, however, match perfectly with a KLV camp: a house filled with militarized children aged ten and under, all fully indoctrinated in Nazi dogma and propaganda.

---

10. https://en.wikipedia.org/wiki/Evacuations_of_children_in_Germany_during_World_War_II.

11. https://www.historyplace.com/worldwar2/hitleryouth/hj-boy-soldiers.htm

Left: German children leaving by train for a KLV facility. Their gleeful expressions belie the grim fate awaiting them.

Right: KLV propaganda poster reads, *"air emergency areas are no place for children! Come to the KLV: A place for Hitler Youth."*

Above. Nancy Miles poses with, from left, Herr Meyer Hofer and a young waitress who served as our interpreter, in a restaurant in Schonegg, Germany. Herr Hofer was eight years old when Ralph and the rest of the 12th regiment arrived in his village on May 3, 1945. He recalled that "he had a very good experience with" the GI's, who gave out lots of candy.

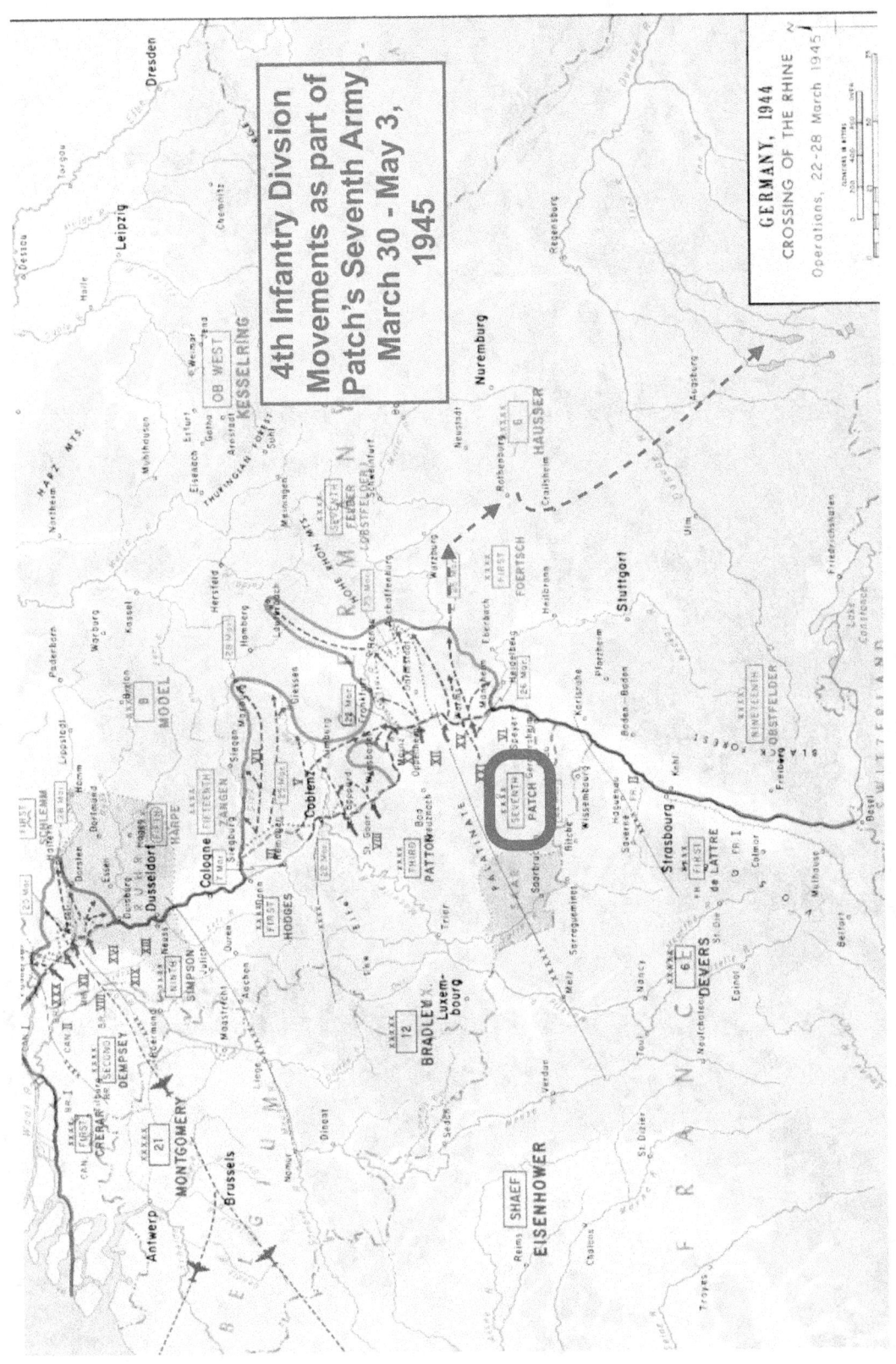

4th Infantry Divsion Movements as part of Patch's Seventh Army March 30 - May 3, 1945

GERMANY, 1944
CROSSING OF THE RHINE
Operations, 22-28 March 1945

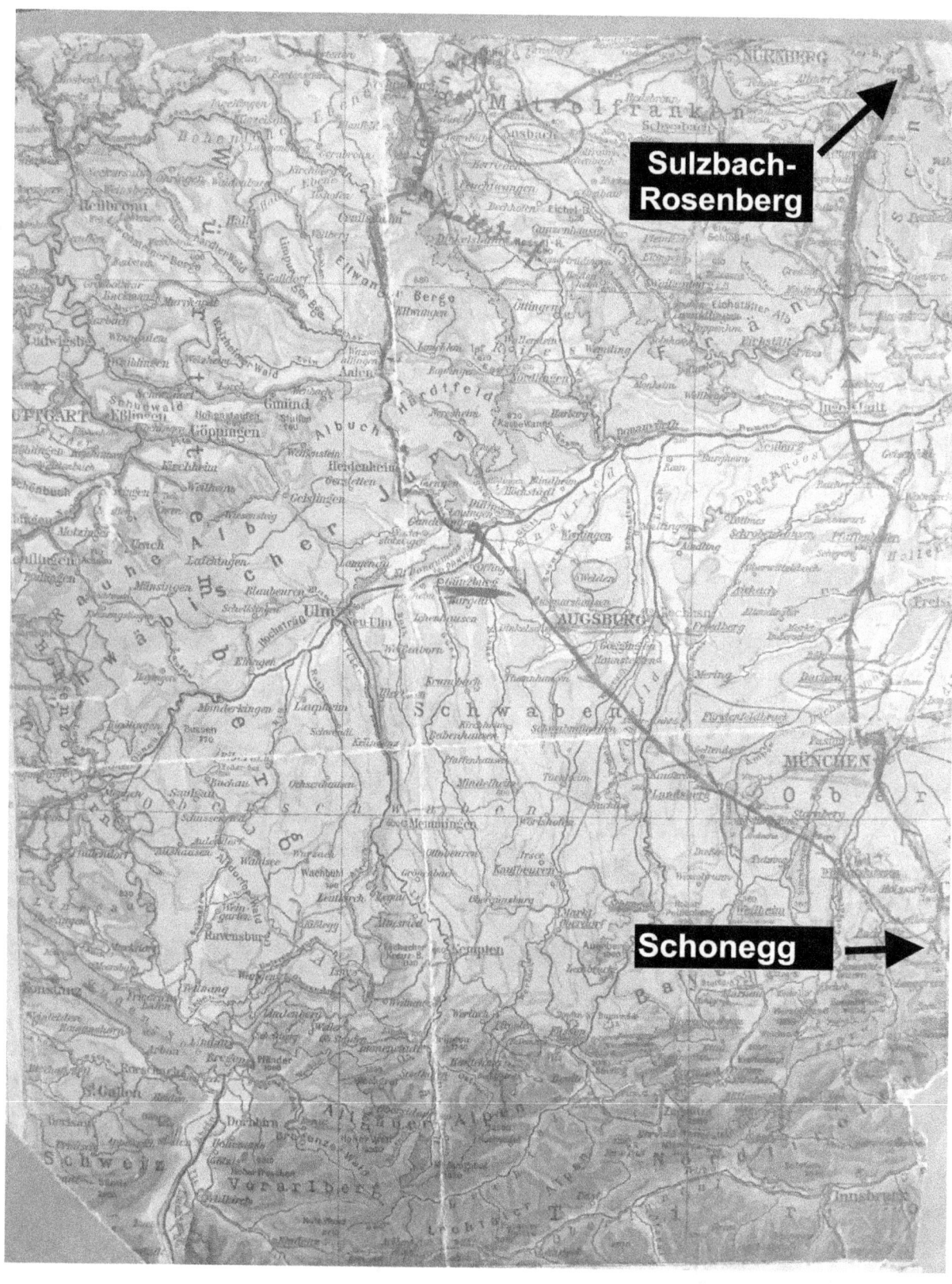

13. The Best of Times, the Worst of Times

"... it was the saddest day of my life to this point."

On May 5, our half-track pulled into Sulzbach-Rosenberg, a place where we were to spend some time. There was a school of kids belonging to the Hitler-Jugend (youth). They were intensely interested in all things military and swarmed over our equipment like so many honeybees. Almost all these kids could speak English. They said they learned it so they would be prepared when the Germans took over England and America. They put no stock whatever in our protestations that the war was over — they insisted such talk was purely propaganda. Hitler would invent more and better secret weapons and, "You'll see!"[1]

May 8, 1945, at Sulzbach-Rosenberg was a great day — the war in Europe officially ended — I kept my copy of the *S&S* announcing the surrender! On the facing page is a map that I tore from some schoolboy's geography book along the way, on which I plotted our travel from Rothenberg to the Alps, back up to our occupation area in Sulzbach-Rosenberg and thence to Windsheim and Bamberg.[2]

We had a pretty impressive place to set up shop when we got to Sulzbach — in the center of the beautiful *Schloss* Sulzbach-Rosenberg. It was an enormous complex and, like everything else in Germany, dated back to medieval times or earlier. Its various wings were arranged in a kind of semi-circle and appeared to be made from white stucco or stone. Our half-track sat just outside the inner plaza on a grassy area that

---

1. These children were living in a *kinderheim* or children's home, in one wing of the Schloss. The parents of these children had either been killed in the war or who had been deemed by the Reich as inadequate in their indoctrination of their children in Nazi dogma. The care and ideological environment of the *kinderheim* closely resembled that of a KLV camp, and the outcome was much the same for the children — there was no place for kindness or tolerance, and any sign of weakness, mental or physical, invited abuse and domination by the stronger children. In a picture Ralph took of one of these boys, you can almost sense in his body language and stance the swaggering cockiness that these camps and homes instilled in their charges. (Source: (1) Thomas Hopper, government engineer and historian at the *Schloss*; (2) https://www.historyplace.com/worldwar2/hitleryouth/hj-boy-soldiers.htm:).

2. The route that Ralph traced on this map shows that he passed east of Nuremberg en route to Sulzbach-Rosenberg. Yet, Ralph wrote earlier (Chapter 13) that their journey to Sulzbach took them through that badly damaged city, which he or a comrade photographed from his half-track. Ralph may have made an error in his narrative about going through Nuremberg May 4-5, or he may have made a mistake on his map. Later, however, he did pass through Nuremberg when they departed Sulzbach for Windsheim on May 18 — perhaps the photos of Nuremberg were taken during that westerly journey. These small inconsistencies are understandable given Ralph's limited perspective at the time and the fact that he was trying to mark their route while "on the fly."

overlooked the village below. I came across a postcard picture of the place, on which I marked the location of our half-track on the castle grounds.

I made some very good friends while in the Army — Technical Sergeant Nunzio Yocca, was the leader of my group, the Regimental Half-track radio operators. Charles "Chuck" Emery was a highly personable young man. Another great guy was Howard Mathison, a native of Manistee, Michigan who had "done time" with me at Ft. McClellan, Alabama. He was wounded and sent home.

My buddy, Bob Weltzein, and I had similar temperaments and world views, likely accounting for the close attachment we developed during our time overseas. I guess it's true that bonds forged in time of war are deep and lasting. When I was transferred to the FBI office in Portland, Oregon, in 1960, I had a chance to go to Seattle and visit with Bob. It was a great reunion. My only regret was that, for one of the few times in my life, I failed to take any photos to memorialize the event.

After a pleasant two-week stay at Sulzbach-Rosenberg, on May 18, 1945, we moved to Windsheim in the vicinity of Ansbach. We were now part of the Army of Occupation. We anticipated we would be here for an extended period, but we were wrong, as we would soon find out.

We had scarcely settled in at Windsheim when the rumor floated around that one man from each company of the Division was being chosen by lot to get a week's pass to the famed French Riviera in southern France on the Mediterranean. I gave the rumor little credence, thinking, "What chance do I have?" Well, I was the lucky choice!

On May 24, we set out by truck and arrived the next day at Luxembourg City, Luxembourg, via Stuttgart and Saarbrucken. It was a rough ride — across country for the most part since most of the bridges were out. But the worst was yet to come — that train ride from Luxembourg on to Nice, France, was a nightmare long to be remembered.

We were issued a Courtesy Pass Card in the event we were stopped and questioned while at the Riviera — thankfully, I never was. In fact, things were pretty laid back, and the war-weary doughs were given a fairly long leash from which to enjoy the sights, sounds, smells, and joys of this beautiful place, dubbed "Heaven" by vacationing GI's — a lot easier to say (and a more accurate description in our minds) than the official Army name for the area: "United States Riviera Rest Area."

There were a few rules we had to abide by, but they weren't too bad: No swimming (the Germans had sent sewer water into the sea along the Riviera by breaking the conduits), no civilian guests above the first floor, no visits to Monte Carlo (a casino), curfew at 1 am. Nothing I couldn't live with, and certainly nothing to keep me from having a wonderfully relaxing few days in a place I never dreamed I would visit in my lifetime. Too bad it took a world war to get me there.

Upon arriving in Nice, we were issued a little booklet that contained all kinds of information for the GI "tourist": a brief history of the region, places to go, movies, plus a few pointers on how to avoid getting into trouble with the MP's while still having a good time. At the back of the pamphlet were some blank pages entitled, "Notes," on which I made this entry:

"Here is a promissory note. I promise to deliver in person, date unknown, 10,000 kisses to one Mary Eleanor Hobbs Miles, in return for which I will expect a like number of kisses.

SIGNED,
Ralph J. Miles"

I may have been the in the romance capital of the world, but my heart was in Savannah, Georgia.

It was beautiful in Nice, more than enough to make up for the nightmare journey to get there. Luxurious is the word. Real food served by French waiters, sheets on the beds, plenty of rest, no saluting, dress as you please. I'm sure I consumed gallons of ice cream. A pleasant way to wind down after a hard war.

One of the things I made sure to do was to document my time in "Heaven" for any doubters back home. After all, it's a long, long way from Tattnall County to the French Riviera. So, I made a stop at a photo booth I had spied on the Riviera and had some pictures made. I also found a street photographer to take my picture. I was no Clark Gable, but I figured at least one Eleanor Hobbs Miles would be duly thrilled at the result.

While in Nice, I was billeted in the Jardin (French for "garden") Prince Albert I Hotel. My room was on one of the upper floors, which gave me a spectacular view both of the gardens below and of the Mediterranean Sea no more than a hundred yards away. We ate in a huge building that sat facing the shore on the wide boulevard that ran along the coast, the Promenade des Anglais.

One day during my stay in Nice, those who cared to go (I did) were driven through famous Cannes to Grasse, a noted perfume-making location. The hillsides were loaded with flowers. I held a card over a vat where perfume was being extracted, infusing the card with fragrance that lasted for years. At Grasse I bought for a small price two bottles of Channel #5, which I brought home for Eleanor.[3]

As they say, "all good things must come to an end," so on June 4, 1945, all of us Riviera beachcombers began our return trip to Windsheim. On the way back, we spent a night in Luxembourg, a city with which I had become somewhat familiar, having been there on leave twice before. I stayed in a building reserved for American GIs on Avenue Marie-Therese.

Upon arriving back at Windsheim on June 7th, I received a cable from my brother Curtis that my mother had died. I tried through every channel to get passage to the States. Even went to see the Chaplain. It was a solid "No" all along the line. I was told the whole Division was returning to the States soon to be refitted and sent to the South Pacific and that I would not be able to return ahead of the Division. "Besides," I was told, "your mother is already buried and there is nothing you can do." It was the saddest day of my life to this point.

On June 9, 1945, we moved out of Windsheim by truck convoy northward to a huge field in the

---

3. The company where Ralph got the card and the perfume, La Parfumerie Fragonard, is still in business in Grasse.

vicinity of Bamberg, Germany.[4] The area immediately became "tent city." Bob Weltzein and I shared a pup tent during our month-long stay in Bamberg. If you have to be cooped up in cramped quarters like this, you might as well do so with someone you get along with. During our time near Bamberg, we took a number of pictures, as one of the guys had a camera, a prized commodity among GI's!

It was a short occupational period for us. The official word soon came: The Fourth Infantry Division was to be redeployed to the South Pacific Theater of Operations via the United States. Nobody liked this idea — least of all me.

---

4. The 12th Regiment Command Post was actually located at Schmerldorf, a village only about one mile northeast of Bamberg. Ralph and the troops encamped in the broad fields between the two towns.

Above: Postcard showing where we parked our halftrack near Sulzbach Castle from 5-18 May 1945

Above: Another postcard showing some of the main buildings in Sulzbach — clockwise from top left: City park, town hall, city pool, parish church, town square.

Robert T. Weltzein
SULZBACH-ROSENBERG, GERMANY
MAY 1945

Of Seattle, Wash.

Charles
Emery
of
Erie, Pa.

SULZBACH-ROSENBERG,
GERMANY. MAY 15 1945

Our casual poses from our time in Sulzbach-Rosenberg speak volumes about our state of mind. The stress of combat had been lifted, and we could finally start thinking about going home and picking up our lives with our loved ones once again. It was just a little matter of an extended Army of Occupation duty first, though … or at least so we thought. (As you can see, while overseas I got "schooled up" in the fine skill of firing up a pipe!). **Clockwise from top left**: Bob Weltzein, Charles Emery, and me.

SOUTHERN GERMANY EDITION
Tuesday, May 8, 1945
Volume 1, Number 1

THE STARS AND STRIPES

Daily Newspaper of U.S Armed Forces · in the European Theater of Operations

Ten days that shook the world — a review of the last ten days, the most important in modern history, appears on Page 6.

ETO
WAR
ENDS

Unconditional surrender of all German forces was announced yesterday by the German radio at Flensburg. Grand Admiral Karl Doenitz, successor to Adolf Hitler as Fuehrer of Germany, ordered the surrender and the German high command declared it effective, the German announcement said. No immediate confirmation came from the capitals of the Allied powers, but President Truman and Prime Minister Churchill will broadcast at 3 p. m. ETO time today, and King George VI will broadcast tonight at 9 o'clock.

In London, the British Ministry of Information announced that today would be V-E Day, and that today and tomorrow would be holidays in Britain. In Washington, President Truman said he had agreed with London and Moscow to make no announcement of surrender until a simultaneous statement could be made by the three governments.

U.S. Celebrates Victory 2nd Time in Ten Days

NEW YORK, May 7—Tension-light America exploded today for the second time in 10 days over the unconfirmed report that Germany had surrendered unconditionally to the Allies.

In New York, clouds of torn paper — Gotham's traditional form of celebration — began swirling through the city's granite canyons a quarter of an hour after the news flash reached the population.

Office workers by the thousands appeared at their high windows, casting out handfuls of paper scraps. In Brooklyn, 38 girl garment workers fainted when the news was flashed. After they had been revived, all 308 employees in the loft were given the day off.

King Expresses Thanks to Ike

New S and S Is Off To a Rousing Start

Here's our new baby—the Southern Germany edition of The Stars and Stripes, which begins publication this morning. And no Volume 1, No 1, has ever got off to such a bombastic send-off as that oculist's eye chart you see topside.

We've opened the new edition—the second in Germany—to improve circulation to the farthermost areas, thus getting the paper out to you in better time. The Pfungstadt edition, which began publication April 8, will continue to service the troops in northern Germany.

Organizing two separate Germany editions will enable us to give you a more complete news-

---

Above: Front page of my copy of the May 8, 1944, issue of *Stars and Stripes* telling of Germany's surrender. Needless to say, we all read every column with a combination of intense interest, great relief, and unrestrained joy. Later we would learn that the war wasn't quite over for us in the 4th Division: after a month state-side to rest and re-fit, we were to ship out to the Pacific to participate in the planned invasion of the Japanese mainland.

Nunzio S. Yocca

WINDSHEM, GERMANY 6-8-45

of Windber, Pa.

howard mathison

of Manistee. Mich.

Robert "Bob" Weltzein

Clockwise from top left: 1) Nunzio Yocca, Technical Sergeant, **2)** Howard Mathison, of Manistee, Michigan, **3)** Bob Weltzein, of Seattle, Washington.

I grew very close to these fine men during our months together in foxholes and riding a half-track in and over the European countryside.

TENT CITY - WINDSHEIM, GERMANY
JUNE 8, 1945

Clockwise from top left: 1) Bob Weltzein and me in front of our tent in Windsheim, **2)** Me, Yocca, German Hitler Youth, Wheelock in Sulzbach-Rosenberg, **3)** L-R: Bob Walters, Jack Verga, Robert Nace, and me.

12th Infantry Radio Section, HQ and HQ Company — Bamberg, Germany

1 — Bob Walters

2 — Jack Verga, The Bronx, NY (1919 - c 2000)

3 — Pfc. Robert F. Nace, Minersville, PA (1915 - 1997)

4 — Half-track radio group leader T/Sgt Nunzio Yocca, Windber, PA (1920 - 2010)

5 — Radio Section Chief T/Sgt Raymond Kimmel, Lebanon, PA (1907 - 1987)

6 — Pfc. Francis L. Wheelock, Hornell, NY (1917 - 1968)

7 — Pfc. Walter R. Kubiak, Chicago, ILL (1918 - 2009 — Landed at Utah Beach)

8 — T/4 Joseph J. Young, Chicago, ILL (1916? - 1999?)

9 — T/5 William D. Harris, Zanesville, OH (1916 - 2010)

10 — Swede Carlson

11 –Pfc. Ernest C. Jarrett, Greensboro, NC (1919 - 2006)

12 — S/Sgt Ralph J. Miles, Senior, Tattnall County, GA (1921 - 2008)

13 — S/Sgt Earl W. S. Gehman, Anderson, NC (1920 - 1986)

14 — T/4 Emil C. Lukert, Philadelphia, PA (1915 - 1972)

15 — Pfc. Howard Meese, Greensboro, PA (1918 - 1989)

Not in picture: T/4 Charles Manton Emery, Erie, PA (1921 - 1987) and T/5 Robert T. "Bob" Weltzein, Seattle, WA (1921 - 1997)

Right: I won the lottery! No money, but a free trip to the French Riviera.

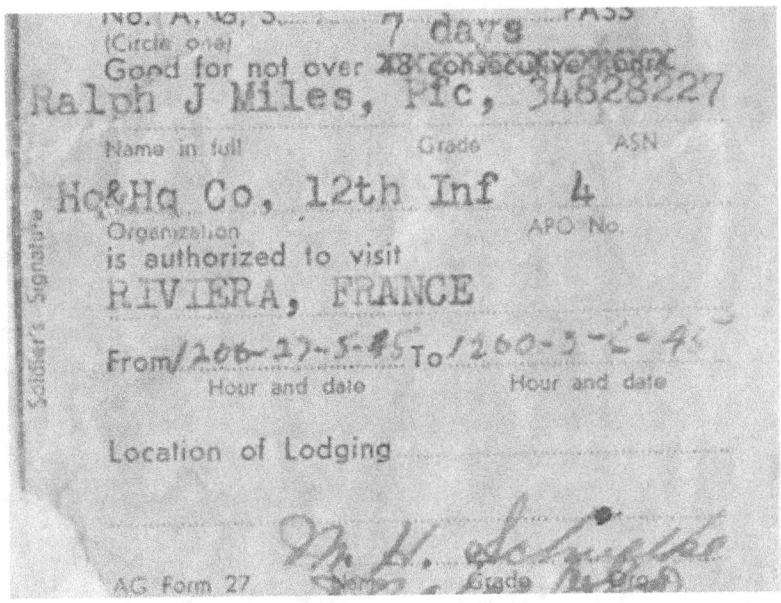

Left: My pass card for my stay in Nice. "USRRA" stands for "United States Riviera Rest Area." We simply called it "Heaven."

Right: The fumes from the vat at La Parfumerie permeated this little card so thoroughly that the aroma lasted for years afterwards.

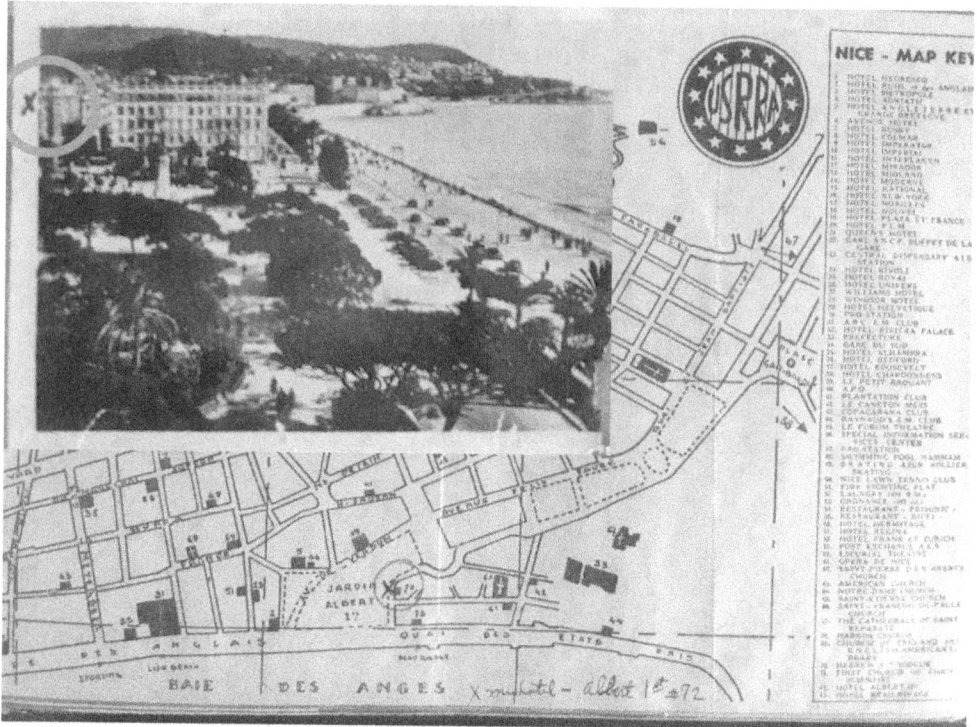

Above: Prince Albert I Hotel (marked by a black "X" inside the circle) where I stayed in Nice.

Left: View of the Riviera from my room on the sixth floor of the Prince Albert I Hotel.

Left: The Red Cross took over a large casino overlooking the Mediterranean Sea to feed the thousands of GIs on R&R in Nice. We ate in a huge ballroom that, before the war, was filled with the world's rich and famous.

Above: Postcard showing in the left background the building where I stayed 4 and 5 June 1945 during my return trip to Bamberg from Nice.

Right: Guidebook that the Army issued to us GI's to help us have a good time without ending up "doing time." I wrote a note to Eleanor in the back just to let her know that she was on my mind, not any of the mademoiselles I encountered on the Riviera!

Pinch me! I'm on the French Riviera! Had to bring home some pictorial proof so the folks back on West 41 Street would believe it!

Above: Me standing on the Promenade des Anglais in Nice.

* * * * *

Left: Modern aerial view of Schloss Sulzbach-Rosenberg showing where Ralph's half-track squad parked from 5 -18 May 1945.

A view of the parish church in Sulzbach-Rosenberg from the same angle as the picture on the postcard.

Nancy stands in the same spot as the GI seated on the wall (in circle) in the left background of the photo of Charles Emery.

The *rathous*, or town hall, stands on the town square directly opposite the parish church. It seems fitting that a place where politicians meet is well, a "rat" house!

Above: This wing of the castle in Sulzbach housed a *kinderheim* for displaced German youth. Here they would be fully indoctrinated in Nazi ideology and propaganda. It was from this building that a swarm of kids eagerly emerged to examine Ralph's radio half-track that was parked on a clearing located at the left end of this wing. This section also served for a time before the war as a prison for women — the windows along the bottom floor are still covered by bars.

Above: Open fields between Bamberg and Schmerldorf — perfect for bivouacking thousands of Ivy Men.

Left: Cobblestone ramp that ascends to the schloss from the village below. The radio section half-track rumbled up this ramp upon arriving in Sulzbach 5 May 1945.

The French Riviera, Nice, France (Clockwise from top left): 1) Modern view of the Prince Albert I Hotel on 4 Max Gallo, 2) Building at 13 Promenade des Anglais, where Ralph and his fellow GIs were fed by the Red Cross — now the Palais de la Mediterranee, a Hyatt Regency luxury hotel (only the façade remains from the original structure), 3) Modern view from a fourth-floor room of the Prince Albert I hotel where Ralph stayed in 1945. The statue visible in Ralph's photo is now obscured by trees, 4) David stands on the balcony of an upper-level guest room in the Prince Albert I hotel overlooking the Riviera, just as Ralph did 80 years earlier, 5) Photo that hangs inside the Palais de la Mediterranee showing an interior ball room as it appeared during the war — Ralph and his fellow doughs would have eaten in one of these rooms.

Above: Enlargement of Ralph's postcard photo of the building on Av Marie-Therese.

Above: Nancy standing in front of the building on 12 Av Marie-Therese where Ralph spent the night of June 5, 1945, while en route to Bamberg from Nice. The interior of the edifice was undergoing extensive renovations in 2023.

Left: Nancy and David stand at the foot of a grand portico on the side of the same building. A portion of the peaked gable on the front of the building is visible here jutting above the roof on the left side.

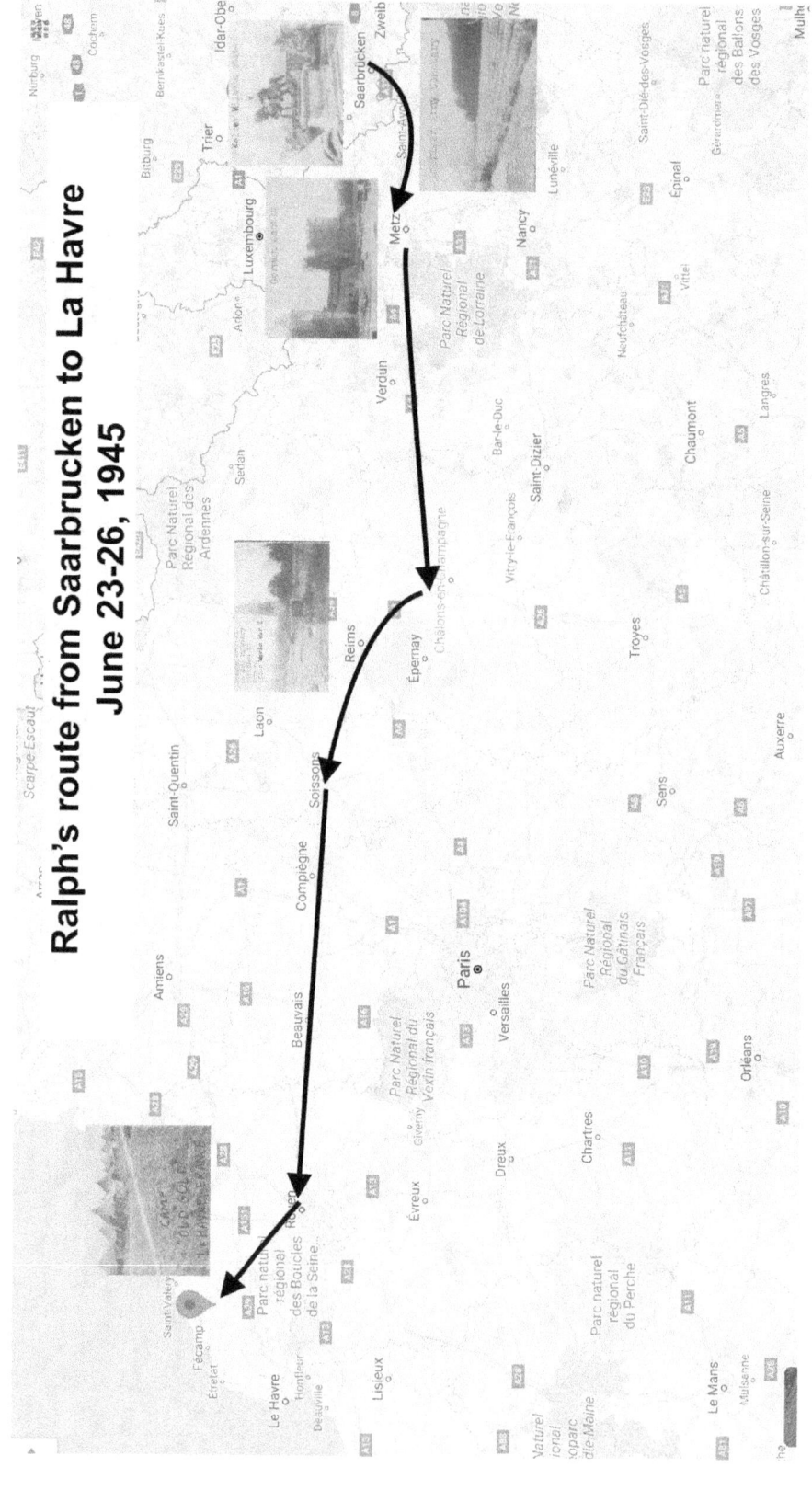

Ralph's route from Saarbrucken to La Havre
June 23-26, 1945

14. Sentimental Journey

"It was roll, heave, upchuck!"

At Bamberg, Germany, some 2,000 men with high point scores toward discharge were transferred from the 4th Infantry Division to the 99th, and a like number of low-point men were transferred from the 99th to the 4th.[1] Because, you see, the 4th was going to finish off the Japs!

Accordingly, in a long overland truck convoy,[2] the 4th Infantry Division moved from Bamberg, Germany, June 23, 1945, and arrived at Le Havre, France, June 26th. I borrowed a camera during our travels and snapped several photos of some noteworthy landmarks we passed by on our route — the most sobering of which were the American Military Cemetery in Saint Avold, just across the French-German border, and a monument to the valor of our division in the Battle of Asne-Marne near Reims, France, in WWI.

Once we finally arrived in the Le Havre area, we were billeted in large tents. Our area was designated "Camp Old Gold." We anxiously awaited our turn to board ship in our "luxury" accommodations in Camp Old Gold, where we turned in our rifles and went through a number of other wind-down preparations before embarking for home.[3]

While waiting at Camp Old Gold to board ship for the United States, I completed a home-made anniversary card and mailed it to Eleanor airmail so it would get there before I did — it may read a bit "sappy," to some, but it expresses exactly how I felt — and still do!

Also during our time near Le Havre, we were given a slip of paper listing the various medals we had earned while in the ETO. As you can see by my handwritten notations, the Army decided sometime after

---

1. The point system used at that time sometimes sent men with fewer than 80 points back home for furlough and then retraining for the Pacific Theater. Points were awarded based on the number of months in the service, number of months overseas, number of children, number of battle stars and individual medals and unit citations won. Ralph had accumulated about 78 points. (source:http://www.skylighters.org/special/cigcamps/cigintro.html)

2. Most of the regiment traveled via rail — Ralph was among the remainder that rumbled along in troop transport trucks and other regimental vehicles (Johnson, p, 349).

3. "Equipment was packed and stenciled, foreign currency was converted into American dollars, Paris passes were granted for the last time to a fortunate few, [and] Barracks bags were checked for unauthorized equipment." (Johnson, p, 350).

the end of the war that I was due a few more tokens of Uncle Sam's appreciation. The four Bronze Battle Stars were for the Hurtgen, the Ardennes Offensive (Battle of the Bulge), our campaign in the Eiffel to take Prum, and our final drive across Central Germany and into Bavaria.

How impatiently we awaited the journey home! After a seemingly endless week, on July 5, 1945, we were put aboard a U.S. Liberty ship,[4] the SS *Sea Bass*. While there was an air of excitement on board, our anticipation was somewhat lessened by the prospect of a few weeks rest and then another sea journey to the Pacific.[5]

All the way across the Atlantic, the ship's loudspeakers blared out over and over again the tune, "Sentimental Journey." Until this day I can never hear that tune without getting a little seasick. The *Sea Bass* was a much smaller ship than the one I had first crossed the Atlantic in. It was roll, heave, and upchuck![6]

July 12, 1945, saw us entering New York Harbor. Every ship we passed in the harbor gave us three blasts on their fog horns as a greeting and salute. It was a gorgeous day in NY Harbor — the fact is, I have no recollection whatsoever about the weather conditions, but it was a gorgeous day nonetheless to all of us home-hungry Warriors!

We unloaded from the *Sea Bass* at Camp Shanks, New York, the same place where I had boarded the HMS *Aquitania* one year earlier, almost to the day. While at Camp Shanks, we were treated to the most fabulous dinner it was ever my experience to enjoy while in the service: cake, chicken, ice cream, milk, and many, many more good things![7]

I was torn between the desire to telephone Eleanor and hear her voice versus the desire to get on home and completely surprise her. The latter choice won out.[8] After only one night at Camp Shanks, our new

---

4. Nearly 3,000 Liberty ships were built during the war as cargo vessels and, starting in 1943, as troop transports. They were not known for their beauty or sleekness (FDR pronounced them "dreadful looking objects"). (https://en.wikipedia.org/wiki/Liberty_ship). Ralph said they were made of concrete, which always greatly taxed my imagination — just couldn't envision a ship made of concrete — at least, not one that would float! It was only after recently touring a similar ship now anchored in Tampa Bay, Florida, that I learned that the ships' hulls were indeed concrete; the docent explained that the decks were constantly having to be re-poured due to cracking from the ship's movements as it wallowed and pitched its way through the high seas!

5. A fellow Ivy Man also on board the *Sea Bass*, Staff Sgt Henry C. Strecker of Company C, remembered that they boarded between nine and ten a.m., left the harbor about nine p.m. and "watched the shoreline disappear in the dusk." (Courtesy of Sgt Strecker's daughter, Leslie Weiner, in a post on National 4th Infantry Division Association FB page, July 23, 2021).

6. Sgt. Strecker recalled that a bad storm hit the ship about two days out of New York, and the ship was full of seasick, land-lubbing dogfaces! (Ibid.)

7. Sergeant Strecker's recollections confirmed Ralph's wondrous description of this savory spread: "A feast of steak, all the milk they could drink, peas, gravy, mashed potatoes and ice cream … wonderfully unbelievable real food after all the box rations … " (National 4th Infantry Division Association FB post by Leslie Weiner, July 23, 2020.)

8. Eleanor recounted the story of Ralph's arrival many times in subsequent years. She remembered sitting in the front room of her late

First Sergeant, Sanders Ray of Covington, Georgia, and I entrained for Ft. McPherson in Atlanta, Georgia, where I had mustered in fourteen months — a seeming lifetime — before.

Needless to say, that thirty days was a joyous time as I got re-acquainted with Eleanor and daughter Sharron, and experienced the incredible thrill of meeting my son Ralph Junior for the first time. During this blissful period, war's dread fears and the long separation were forgotten, swallowed up in the inexpressible bliss of reunion.

During my furlough in Savannah, I made sure that we spent a day at one of my favorite places, Savannah Beach on Tybee Island. Needless to say, I had a marvelous time — it was so good to feel fine sand once more between my toes instead of cold European mud, to again experience the warmth of the southern sun on my face rather than the sting of sleet and snow, and to cavort with wife and children in the bath-like water of the Atlantic Ocean instead of shivering in some ice-water-filled hole in the ground.

Happily — for us, anyway — during this furlough, atomic bombs were dropped on the cities of Hiroshima and Nagasaki, and Japan surrendered August 14, 1945. Hope rose that I would not, after all, be shipped to the Pacific.

Alas, my thirty blissful days at home came to an end, and I rejoined my Regiment and Division at Camp Butner, North Carolina. On September 21, 1945, I was promoted from Private First Class to Staff Sergeant. It was nice to get a promotion — but the job that came with it, being in charge of the Regimental Message Center — I thoroughly despised. But the good news was that, with Japan's surrender, the War Department immediately began the process of demobilization, and we all knew we would not be shipped back out and could look forward to our imminent discharge from the Army.

Indeed, much to my relief, and much earlier than I had even dreamed possible, that Great Day came — October 14, 1945 — when it was my turn to go over to "Happy Valley." I cannot describe the feeling I experienced when, at long last and after so many trials and dangers, I was now free to return to my beloved Savannah and to my small but growing family ... this time, for good.

---

mother-in-law's home and then hearing footsteps on the porch. She ran to the door and flung it open to reveal Ralph's beaming face and trim GI physique, and, well you can imagine the tearful, unrestrained joy of the embrace that ensued!

I borrowed a camera from one of the guys and took these photos as our convoy moved across the Saar and through France to Camp Old Gold outside Le Havre.

Right: Statue of Kaiser Wilhelm that stood on the *Alte Brucke* ("Old Bridge")that spans the Saar River at Saarbrucken.

Left: WWI 4th Division memorial near Fismes, France.

Right: Metz, France

Left: As we passed through St. Avold, France we encountered this solemn reminder of who the real heroes were in the war.

Left: Our latest — and last — "tent city" home in the fields northeast of the harbor of Le Havre.

Le Havre, France
June 29, 1945

My Darling Wife,

How well I remember the afternoon of today four years ago when you came to me radiant as an angel. It was the greatest single event of my life, and I hope you will never have cause to regret the pact we made.

My wish, dearest, is that our next anniversary and all that follow will be spent together. I love you.

Always yours,
Ralph

ANNIVERSARY
Eleanor

Above: Card that I fashioned for Eleanor on the occasion of our fourth wedding anniversary.

HEADQUARTERS
12TH INFANTRY
APO 4, U.S. Army

8 July 1945.

Pfc Ralph J. Miles   ASN 34828227 is authorized to
wear the following decorations and ribbons:

~~Silver Star~~ _____ ; Good Conduct Medal ✓ LATER EARNED

~~Bronze Star~~ __ LATER EARNED ; American Defense Ribbon ✓ LATER EARNED

~~Purple Heart~~ _____ ; ETO Ribbon X

~~American Theater Ribbon~~ ✓ LATER EARNED ; Combat Infantrymans Badge X

Unit Presidential Citation __ X __   (Line out decorations not applicable)

Bronze Battle Stars __ 4 __ Arrowhead _____

EARL C. KURTZ, JR.
Capt. 12th Infantry
Adjutant

Above: While at Camp Old Gold, the Army gave us this form telling us what medals we had earned during our time in service. I was awarded Good Conduct, Bronze Star, American Defense Ribbon, European Theater of Operations Ribbon, American Theater Ribbon, and the Combat Infantryman's Badge. All of us in the 12th received the Presidential Unit Citation for valorous actions during the Bulge in defense of Luxembourg City. The four Bronze Battle Stars were for Hurtgen Forest, Northern Germany Campaign (Prum), the Ardennes (Battle of the Bulge) and Central Germany Campaign.

Left: While flipping through my copy of *Life* Magazine sometime in the 1980's, I encountered this Army ad showing the 12th Regiment steaming into New York Harbor on July 12, 1945. Peering more closely at the faces in the picture, I was stunned to find among the rejoicing GI's none other than my own cheering countenance!

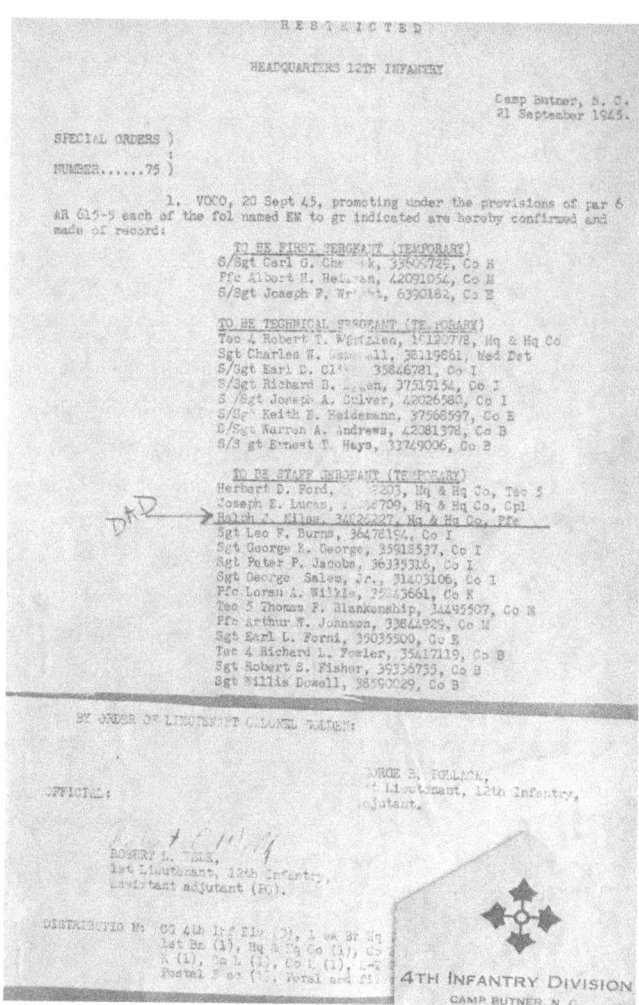

Left: Special Orders Number 75, showing the various promotions while the Regiment was at Fort Butner, NC. I was made a Staff Sergeant and put in charge of the message center — worst job I ever had! My buddy Bob Weltzein was promoted to Tech Sergeant.

Above: Eleanor (holding Ralph Junior) and me (holding Sharron), in Savannah, Georgia.

Left: Eleanor, Sharron, me and Ralph Junior on Tybee Island, Georgia during my 30-day furlough in July and August of 1945. This was then and still is today "my happy place."

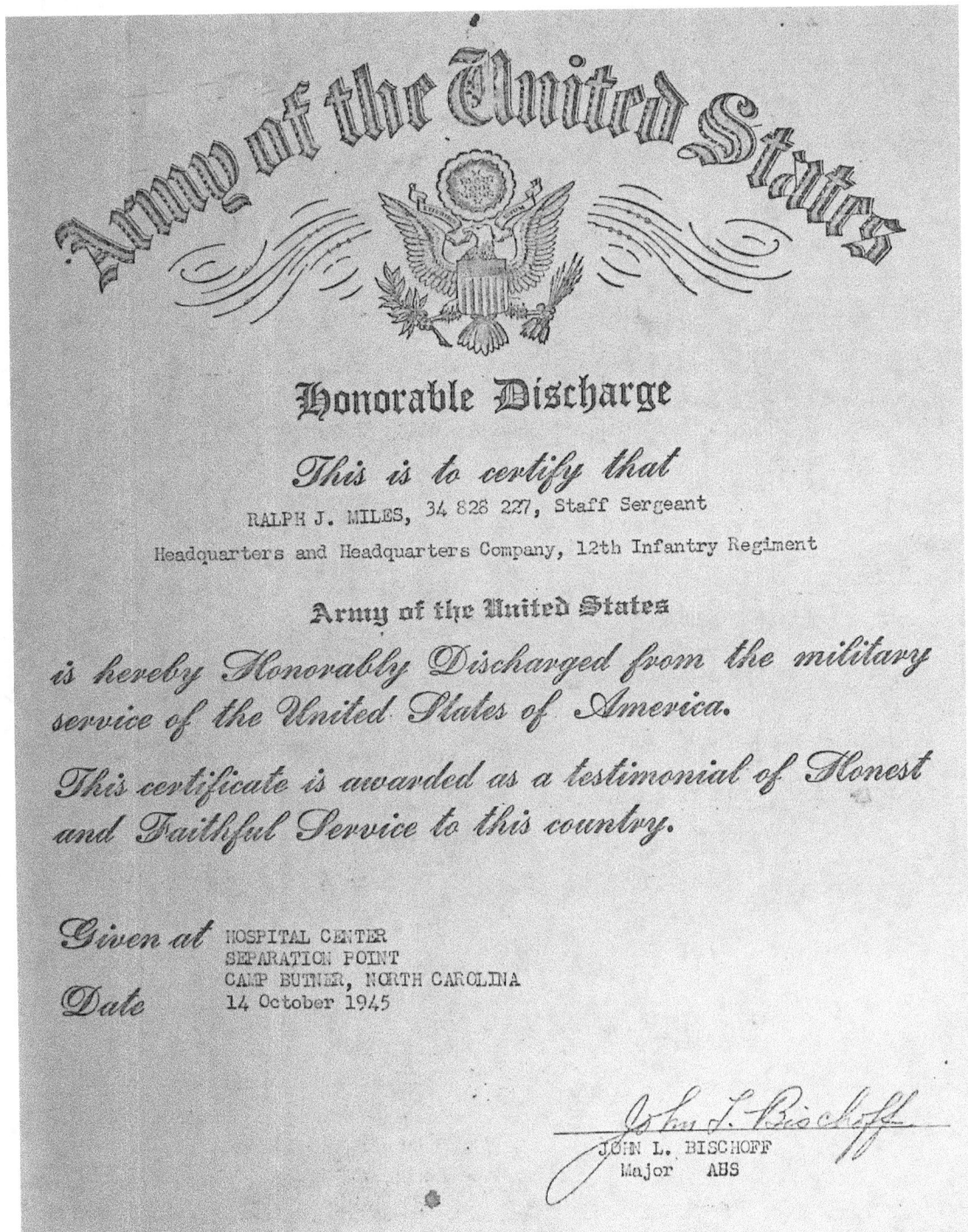

While I had no serious regret about my military service, it was a genuine pleasure to be a civilian again!

* * * * *

Below: Red Warriors anxiously await the "all aboard" signal while on the docks of Le Havre in July of 1945.[9]

USA or bust!

Right: July 5, 1945 — Members of the 12th Regiment (is Ralph in that line somewhere?) board the USS *Sea Bass.*

Left: Crazy camera angle or a pitching and rolling ship deck?? The Sea Bass pulls away from Le Havre, France, and sets out on a Sentimental Journey for the good ole' USA.[9]

---

9. http://www.skylighters.org/special/cigcamps/cmpoldg.html

SS SEA BASS

GENERAL ORDERS

1. SAFETY REGULATIONS

   (a) Life jackets must be worn or carried at all times while the ship is at sea. During emergency drills all life jackets will be kept securely tied. Do not use your life jacket as a cushion or pillow.

   (b) Smoking is prohibited in troop compartments, in the heads, during emergency drills, in the troop mess and galley, and during embarkation and debarkation.

   (c) Smoking is permitted on the weather decks except during blackout, and in the troop mess from the time the galley is secured in the evening until 2300.

   (d) The spaces marked "out of bounds" are restricted to all passengers. The spaces marked "officer country - out of bounds" are restricted to enlisted passengers. Other spaces restricted to all passengers include midship housing, chain room, radio room, sick bay and hospital, permanent crew's quarters and mess, gun platforms, magazines, gear lockers, fan tail, forepeak, lifeboats, rafts, booms, masts, galleys, storerooms, and unoccupied troop compartments.

   (e) Fresh water must not be wasted. Use water from the scuttlebutt for drinking only. If an excessive amount of water is used rationing will be necessary.

2. SECURITY REGULATIONS

   (a) Blackout is announced each evening over the public address system and is in effect until sunrise the following morning. All lights visible from the outside of the ship shall be extinguished during blackout.

   (b) No garbage, trash or article may be thrown overboard except at times designated by proper authority.

   (c) Cameras, flashlights, radios, electric razors, and all other electrical appliances must be turned into the troop office before sailing. This equipment will be returned prior to debarkation.

3. CONDUCT

   (a) The following is prohibited:
   Gambling, loitering, profanity, excessive noise, defacing or removing of ship's gear, sitting on rails or exposed positions.

   (b) Alcoholic liquors shall not be admitted or used on board except as authorized for medical purposes.

   (c) Sunburn is considered misconduct. All troops must be fully clothed at all times.

   (d) Passengers are not permitted to take mattresses, blankets, or bedding from their cabins or compartments.

4. SANITARY REGULATIONS

   (a) Do not throw refuse on the decks, into toilet bowls, urinals, or showers. Sufficient numbers of receptacles are distributed throughout the ship to receive all refuse.

   (b) Food from the galley is not allowed in troop berthing spaces.

Above: General Orders issued to all GI's as they boarded the *Sea Bass* — they knew all about muddy foxholes, stinking slit-trench latrines, frostbite, and trench foot ... but a ship was a bewildering and strange place! (Courtesy of Leslie Weiner, whose father was on board with Ralph and brought this sheet home as a souvenir).

Camp Old Gold was one of the so-called "cigarette camps" erected in the outlying areas between Le Havre and Rouen (plus one camp in Belgium) following the capture of the former on September 12, 1944. Originally built as receiving areas for arriving troops, by the end of the war these sprawling tent cities served exclusively as encampments for troops waiting to board ships in Le Havre bound for the States. The camps were named for American cigarette brands — Chesterfield, Philip Morris, Lucky Strike, Pall Mall, Twenty Grand, Herbert Tareyton, Home Run and Wings — to keep their location a secret from Germans eavesdropping on American radio traffic.

By V-E Day, the soldiers living in the cigarette camps were mostly destined for redeployment to the Pacific Theatre following some time state-side for rest and refitting, as in the case of the Fourth Infantry Division. Very little visible evidence remains of these enormous, muddy, and less-than-hospitable canvass cities, and no trace at all remains of Camp Old Gold, where Ralph and his comrades anxiously awaited their turn to board ship for home.[10]

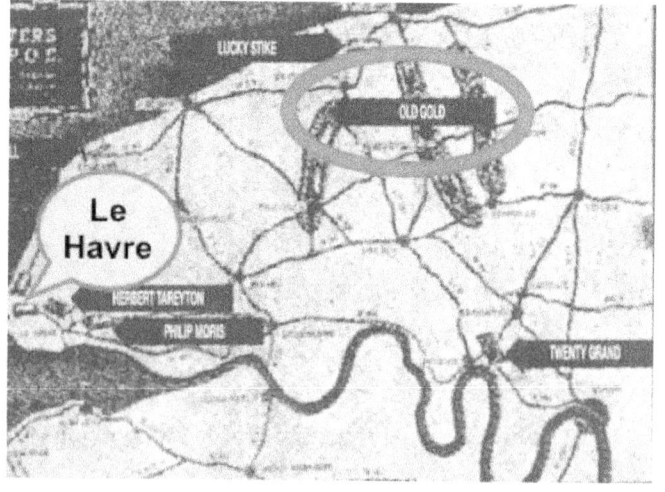

---

10. http://www.skylighters.org/special/cigcamps/cmpoldg.html

Left: Open fields southeast of Ourville, France where Camp Old Gold was located — about 30 miles northeast of Le Havre. Ralph and his fellow Red Warriors languished here for nearly two weeks awaiting their turn to board ship for home.

Left: A muddy "street" in Camp Old Gold.

Right: Troop convoy lined up outside Saarbrucken, Germany, awaiting their turn to cross the Saar River into France. Ralph's convoy followed this same route in similar vehicles three days later — June 23, 1945.

Saarbrucken, International Boundary between France and Germany, June 20, 1945. Column en route to Le Havre.

Lorraine American Military Cemetery in St-Avold, France ("American Army cemetery")

Left: The "American Army cemetery" that Ralph photographed during his trip to Le Havre is located in St. Avold, a village situated west of Saarbrucken and across the French-German border. It is now known as the Lorraine American Military Cemetery.

Right: The picture of the "German castle" that Ralph took proved challenging to identify and locate, for it isn't in Germany and it isn't a castle, though it certainly looks like one. The structure is actually the 13th century *Porte des Allemandes* ("German Gate"), located in Metz, France. This photo was taken from the same vantage point from which Ralph took his picture as his truck rumbled through the city and rounded a curve. Many thanks to Beth Reiman for her indispensable help in identifying the Porte des Allemandes.

Right: David poses beside the 4th Division WWI monument photographed by Ralph in June, 1945. It stands on the side of the roadway west of Fismes, France, on Highway E46/N31. The obelisk honors the Ivy Division for its actions in the Battle of Aisne-Marne in WWI. Members of the 4th Infantry Division Association cleaned and repaired the monument and conducted a flower-laying ceremony on November 11, 2018, the 100th anniversary of the end of WWI.

Above: Aerial view that shows the direction of the regimental convoy (arrow) and the location of the statue of Kaiser Wilhelm. The photo was taken looking backward as the convoy crossed the Saar River on the *Alte Brucke*. **Below:** Ralph's photo superimposed to show its location on the plaza near the *Alte Brucke*. The statue was torn down in 1946.

15. War Relics Photo Gallery

"This was recovered from the body of a solider whose only other equipment was a rifle..."

Over the years, Ralph's children, grandchildren, and countless others have marveled over and handled ever-so-carefully the irreplaceable relics contained in this photo gallery.

I cannot count how many hours I played "Army" wearing one or the other of the German helmets Ralph had "liberated," nor how often my imagination drew vivid mental pictures of the former owners of these items — and the circumstances in which Ralph "relieved" those owners of things of which they would have no more need.

A survey of militaria sites on the web reveals that a number of these artifacts would bring a fair price on the collectors' market. But I'll never see any financial profit from them because no one could possibly afford my price, for their true value is measured not in dollars or euros, but in the blood, love, courage — and lives — given in defense of our greatest national treasure: Freedom.

They are included in this little volume not only as remarkable — and often quite rare — historical curiosities, but also as tangible reminders of the unpayable debt owed by us all to the Greatest Generation.

Medals, Pins , Flags, Patches, Equipment

The Nazis held several annual celebrations intended to build popular support for the Reich. During these celebrations, the Germans distributed what came to be known as "tinnies." These tinnies were small tokens made of tin or aluminum, imprinted with slogans and images befitting the occasion.

Ralph found (or confiscated) two such tinnies. One — pictured on the next page — was the ***Tag der Arbeit,*** created in honor of German Labor Day each May. Each year's badge had its own unique design; this one, as the inscribed date shows, is from May Day, 1935. Etched on the back is the name of the company that produced the badge, *"Hofstetter Bonn."*

Ralph also picked up the tinnie on the right, issued in 1942 during the annual celebration of German Police Day. Like the May Day tinnie, this badge was either given away or sold as a fundraiser for the police department. Many thanks to historian Dr. Jared Frederick for his help in identifying this artifact.

Above left: Ralph uncertainly captioned this artifact *"Iron Cross?"* It is indeed an ***Eiserne Kreuz*** — "Iron Cross." First awarded in 1813 by Prussian Emperor Frederick Wilhelm II for valor in the face of the enemy, the Cross continues be a coveted honor for bravery in the German military.

The soldier from whom Ralph took this medal had apparently served the Fatherland with distinction some thirty years earlier, as it dates from World War I — a fact indicated by the year "1914" etched on the bottom of the cross, by the "W" in the center that refers to Kaiser Wilhelm II, and by the imperial crown engraved at the top. Crosses earned by soldiers during Hitler's Reich have "1939" engraved at the bottom and a swastika in the center.

The Cross could be worn in either of two ways: pinned on the front of the uniform or hanging from a ribbon attached to a loop at the top. The Cross Ralph obtained has what appears to be the remnant of a loop attachment at the top and therefore would have been worn with a ribbon.[1]

The triangular patch on the right below is a **German Luftwaffe M-35 Sports Shirt Eagle.** In addition to the usual military uniforms, the Germans also issued each inductee a full set of athletic gear, including shorts, tank-top style shirts and swim trunks. Each branch of service had its insignia imprinted on its athletic clothing. A larger symbol — like the one Ralph confiscated — was sewn on the back of the tank top, and a smaller version was sewn on the front, as illustrated in the photo at left from the Gettysburg Museum of History.[2] The "M-35" designation indicates that this design made its first appearance on July 14, 1935.

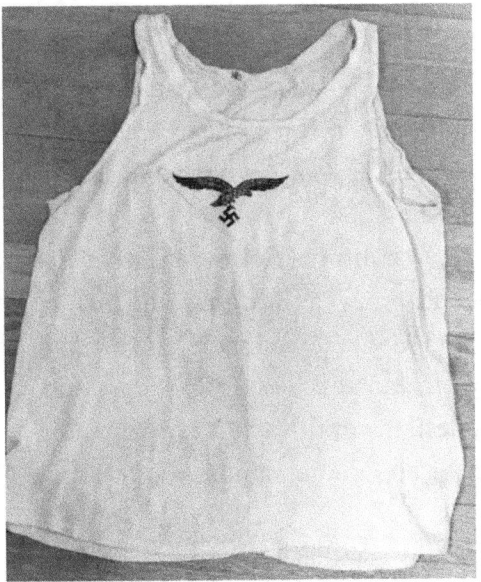

---

1. https://www.britannica.com/topic/Iron-Cross

2. https://www.gettysburgmuseumofhistory.com/gettysburg-battle/world-war-ii- militaria

Ralph may have taken some pleasure in relieving the owner of this particular item — it is a lapel membership pin of the *Nationalsozialistische Deutsche Arbeiterpartei* — National Socialist German Worker's Party, or NSDAP. We, of course, know the Party by its more common moniker, *Nazi*. The wearer wasn't simply some poor soul drafted by the German state, handed a rifle, a helmet and an armband and sent to the front to die. He was a loyal, proud, card-carrying Nazi.

Beginning in 1934, all NSDAP pins were manufactured under the watchful eye of the RZM — *Reichzeugmeisterei* (National Equipment Quartermaster) — to ensure consistency in the design and quality of the materials used. This pin passed muster, as it bears the "*RZM*" stamp inside a circle at the bottom of the reverse side (**above, right**) — the remnants of the pin attachment can also be seen in the picture. The "6" visible at the top of the reverse side indicates the name of the designer of this pin (there was more than one designer): Karl Hensler, whose manufactory was in the city of Pforzheim, famous for its clothing accessories production.[3]

This pin would bring a modest sum on the militaria market — perhaps as much as a hundred dollars.

---

3. https://www.ima-usa.com/products/original-german-nsdap-party-enamel-membership-badge-pin-by-karl-hensler-rzm-6?variant=31668433158213

Above: Nazi flag that Ralph "liberated" from a warehouse in Gundelfingen on April 25, 1945. Note Ralph's notation in the upper right corner: "April 25, 1945. Town of Gundelfingen Germany. Ralph J. Miles"

Ralph translated the inscription on this red and black ***Deutscher Volkssturm Wehrmacht*** armband as, **"German People's Army."** A fuller translation would be, "German People's **Storm** Army." The *Volkssturm* militia consisted of men whose age and/or health exempted them from Home Guard duty, but by the final year of the war, these same men found themselves on the front lines as infantry replacements. Ralph likely picked up these armbands about one mile southeast of Schnelldorf at Volkershausen where the 12th IR encountered *Volkssturm* units on April 20 as Task Force Rodwell was about to jump off on its mission.[4]

Ralph speculated that the yellow ***Deutsche Wehrmacht*** (***above***) armband may have been issued to combatants who lacked proper uniforms late in the war, noting that the armband and a rifle were all he found on the "soldier's" body. In reality, this armband was initially issued to German civilians and, in occupied territories, to foreign workers in the employ of the German military. But as manpower shortages late in the war grew more and more acute, it appears that at least some of these civilian workers were sent into combat — identifiable as soldiers, as Ralph wrote, only by their armband and a rifle. Each branch of the service imprinted its symbol on the armbands of the civilians in its employ. The imprint on this armband reads, "*Kreigsmarine*" ("Navy"), indicating that this individual had been employed at some point at a German Naval base … only to die fighting in a hopeless ground war.[5]

---

4. 12th Infantry Regiment Unit Report No. 319, 21 April 1945 [304-INF(12)]

5. https://www.1944militaria.com/Original_German_WWII_Deutsche_Wehrmacht_Armband_p/ordwamb.htm; q.v. https://epicartifacts.com/product/deutsche-wehrmacht-armband-2/

Above is a detail from the picture **below** of the corpse of *Volkssturm Bataillonsfuhrer* Walter Donicke, who committed suicide in Leipzig on April 19, 1945, to avoid capture by the Allies.[6]

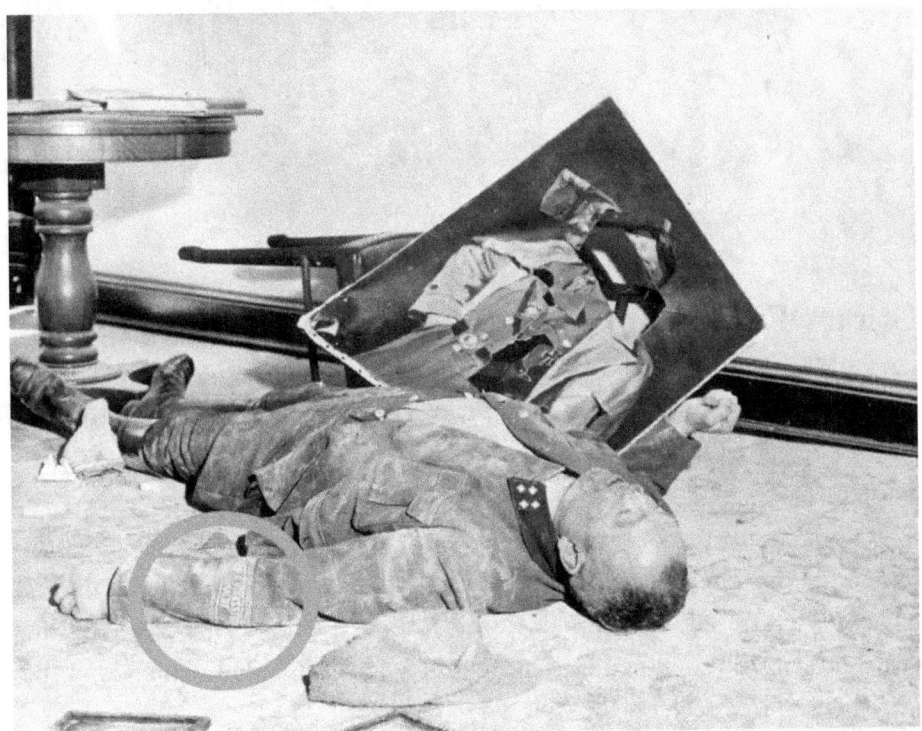

6. https://en.wikipedia.org/wiki/Volkssturm. https://de.wikipedia.org/wiki/Walter_Dönicke

Ralph wrote that the *Wehrmacht* (Army) emblem (**below**) was worn over the "left breast pocket." However, such insignia were always worn on the right side of the uniform. I suspect that Ralph simply got his left and right reversed as he recalled looking down at the slain soldier's body, from which perspective the patches appeared to be over the left pocket.

GERMAN AIR FORCE - LEFT BREAST POCKET

Military Equipment from the Battle for Prum, Germany

In late February through early March of 1945, Ralph and the rest of the Ivy Division spearheaded the assault on and capture of Prum, Germany — known more commonly in history books as the Fight for the Rivers.[7] (see Chapter 8).

On March 2, Ralph and the rest of HQ and HQ company entered the city where they set up the 12th Regimental Command Post. The half-track radio group billeted in one of the few buildings still standing. The room they chose was littered with human refuse and dead Germans. The GI's cleaned up the mess and tossed out the bodies — but before doing so, Ralph kept a couple of mementos of the event: A Luftwaffe helmet and a belt buckle, seen on the following pages.

Hitler pulled members of the Luftwaffe out of their planes and put them on the battlefields as infantry replacements due to mounting casualties in the Wehrmacht.

---

7. Johnson, p. 313.

Above and below: Luftwaffe helmet recovered by Ralph at Prum, Germany, along with Ralph's tag describing the relic.

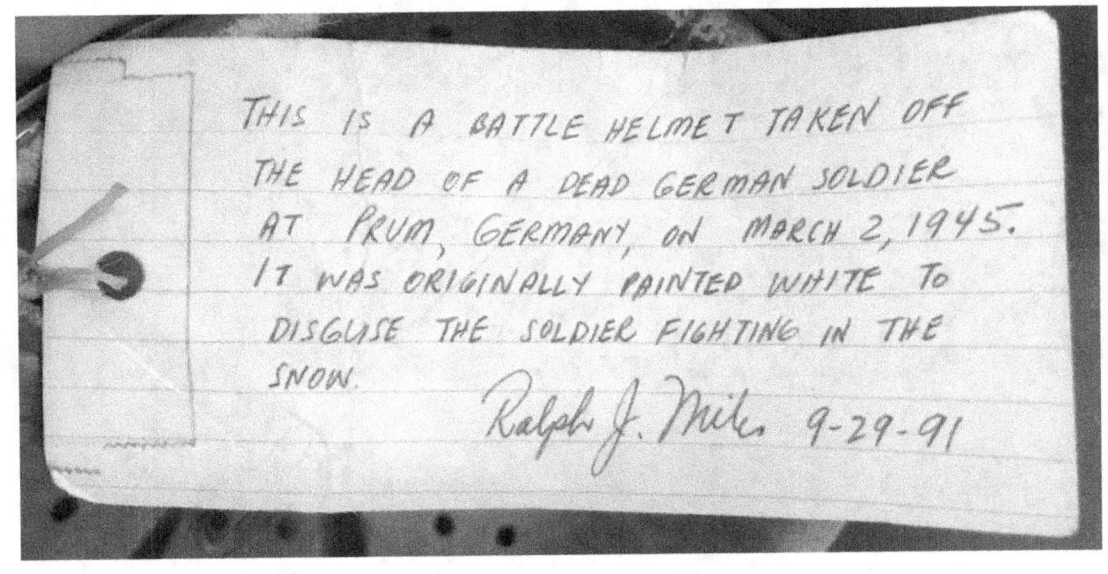

THIS IS A BATTLE HELMET TAKEN OFF
THE HEAD OF A DEAD GERMAN SOLDIER
AT PRUM, GERMANY, ON MARCH 2, 1945.
IT WAS ORIGINALLY PAINTED WHITE TO
DISGUISE THE SOLDIER FIGHTING IN THE
SNOW.

Ralph J. Miles 9-29-91

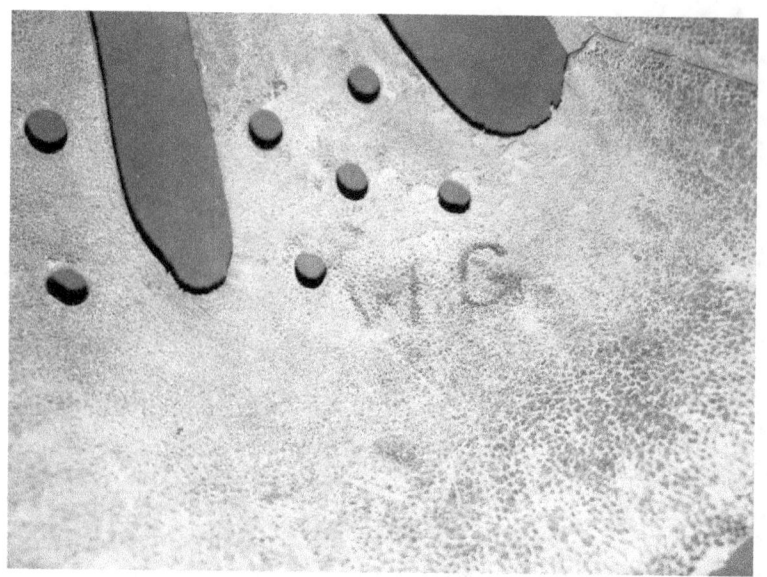

The liner inside the Luftwaffe helmet bears two hand-written inscriptions: "*W.G.*" (**left**) and "*Tim Berardelli*" (**below**). Efforts to date to identify either of these individuals have proved unsuccessful. No one by either this name or these initials has turned up on German rosters so far. This is another of those mysteries that I should have asked my father about when I had the chance, but was too immature and short-sighted to do so.

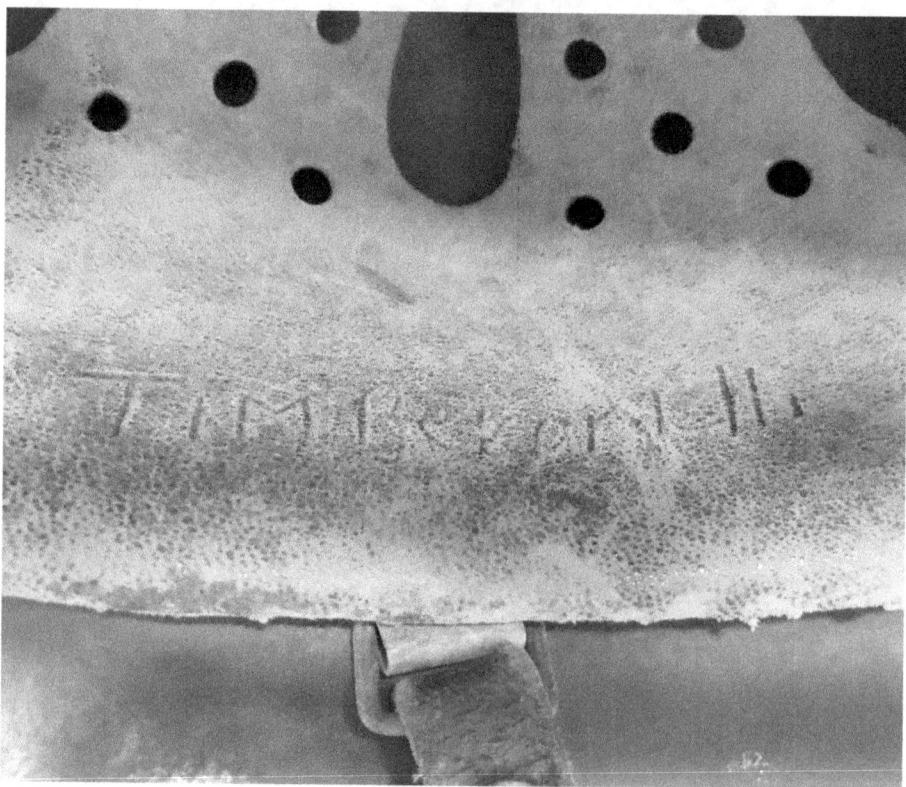

The yellow trim on the collar on the left **above**, indicates the Luftwaffe as the branch of service, and the "*Gr*" abbreviation refers to a *grupe*, the German equivalent of a Wing in the US Army Air corps. On the right is a uniform patch taken from a member of the Glider troops — or so Ralph thought.

Both the collar and the uniform patch belonged to a member of a para-military air force called the ***NSFK*** — *Nationalsozialistisches Fleigerkorps,* or **"National Socialist Flyers Corps."** Established in 1937 by the Nazis, the NSFK continued to operate after the re-activation of the Luftwaffe, functioning mainly as a para-military air-defense organization. Early on, the NSFK trained with gliders, possibly accounting for Ralph's identification of the patch as belonging to "glider troops."

There were seventeen NSFK Groups ("*Gruppen*"), each headquartered in a different region of Germany. *Grupe 15* was based in Schwaben, a region in Bavaria situated between Munich and Stuttgart.[8] Not coincidentally, Ralph and the 12th Regiment passed through Schwaben southwest of Munich in late April of 1945 (see Chapter 12). During this time, the Germans were surrendering in droves to the fast-moving Fourth Infantry Division. It is likely that Ralph liberated this collar either from a surrendered NSFK member pressed into combat duty — or from one who had not surrendered soon enough and suffered the deadly consequences.

According to Olaf Nitsch, this NSFK collar can bring a considerable sum on the WW2 militaria market. Of course, to me it is priceless, like all the rest of Ralph's memorabilia.

8. https://military.wikia.org/wiki/National_Socialist_Flyers_Corps. Many thanks also to Albert Trostorf, Olaf Nitch and Jared Frederick for their invaluable help in identifying these artifacts.

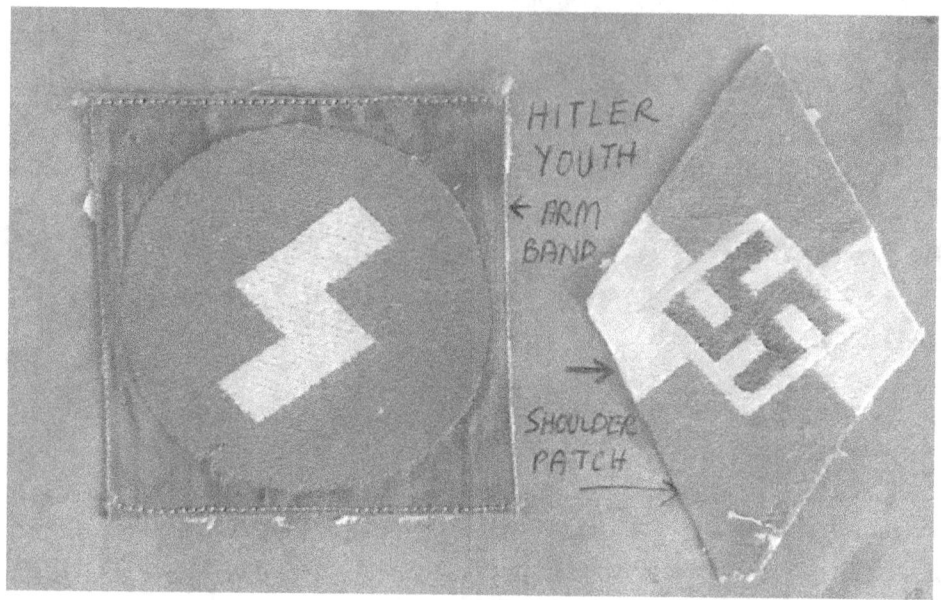

Ralph probably obtained these Hitler Youth patches (**top**) during his two-week stay at Sulz-bach-Rosenberg, where he and his buddies encountered a contingent of Hitler Jugend, and where he photographed one of them eagerly examining their halftrack (**above, left**).

The patch with the stylized "S" on left in the picture at the top of this page was worn on the left shirt sleeve and is known to collectors as the ***Seig Rune Round*** patch. It was the official Hitler Youth "logo" and also adorned belt buckles and banners. This old runic character stands for the "*s*" in "*Sieg*," which means "victory." The *Seig Rune* patch was not given out freely — it had to be earned. The swastika patch pictured at the top right was worn on the right shoulder.[9] A group of Hitler Jugend wearing these emblems can be seen in the photo above right (white arrows).

This last patch (**below**) is especially intriguing, as much for what we don't know as for what we do. We do know that it is a ***German Luftwaffe Qualified Radio Operator's*** patch. It was worn on the lower left sleeve by both non-commissioned officers and enlisted men.[10] Hence, the original owner of this patch was a radio operator just like Ralph! If Ralph took this from a POW, one would love to know what they might have had to say to each other as fellow radio guys.

9. https://histclo.com/youth/youth/org/nat/hitler/uni/insig/ip-bolt.htm.

10. https://www.warstuff.com/product/german-ww2-luftwaffe-qualified-radio-operator-trade-patch/

Historical Coins and Currency

All these coins date variously from the Imperial era (WWI and before), the Weimar Republic (1918-1933) and the Nazi era (1933-1945). Interestingly, the German government did not begin imprinting the Nazi swastika on coins until 1939, some six years after Hitler seized power, which explains why the 50 reichspfennig coin minted in 1935 does not bear the Nazi emblem, while the other coins with later dates do.

**Pre-WWI 10 Pfennig
1907
Note Royal Crest for
Kaiser Wilhelm
Copper and Nickel**

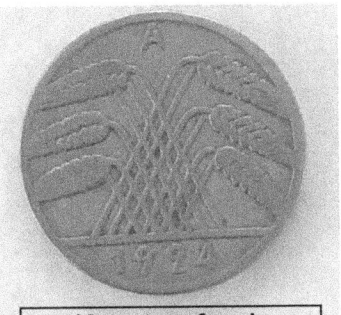

**10 rentenpfennig
1924
"A" = Minted in Berlin
Aluminum and bronze**

**5 reichspfennig
Weimar Republic coin
1925
Aluminum and bronze**

**1935 50 Reichspfennig
"German Empire"
Obverse side: "A" =
Minted in Berlin
Zinc** 21-25

**5 reichspfennig
1939
Obverse side: "B" =
Minted in Vienna, Austria
Bronze**

**1 reichspfennig
1941
"A" = Minted in Berlin
Zinc**

**10 reichspfennig
1944
"F" = Minted in Stuttgart
Obverse side:
Nazi eagle and swastika
Zinc**

Above: This 100 reichsmark banknote was authorized before the start of WWI by the Imperial Bank in 1903 during Wilhelm's reign. It maintained its face value until the catastrophic hyperinflation of the post-war period.[11] As a collector's item it would be worth a considerable sum today — but only if it had a six-digit serial number, meaning that it was printed before the inflationary time. Sadly, Ralph's *"souvenir German inflation note"* has one digit too many!

On the following page is another sample of Wilhelm-era currency, this one initially authorized April 21, 1910. The thousand-mark note, according to Ralph, would have been worth 100 USD under normal circumstances. Before the start of WWI, however, it had a USD equivalent of $238; by war's end, it had dropped to $142.[12] Of course, it has long since been de-monetized, like all the samples in Ralph's collection, and is worth no more than the paper it is printed on. All these notes bear the Imperial Eagle in red, denoting that their authorization occurred before the fall of the Empire in 1918.

---

11. The inflation rate in Weimar Germany reached an astounding 29,500% by the fall of 1923. (Fischer, Wolfgang C., ed. (2010). German Hyperinflation 1922/23: A Law and Eonomics Approach. p. 91, quoted in https://en.wikipedia.org/wiki/Papiermark.) A working-class German citizen carried in his pocket bills in denominations of as much 1 trillion reichsmarks in order to buy essential items such as bread, meat and heating coal. (Nick Goodell, https://www.spurlock.illinois.edu/blog/p/1920s-hyperinflation-in/283)

12. https://www.collectorsweekly.com/stories/225099-1000-mark-reichsbanknote-imperial-germ; also https://www.leftovercurrency.com/1000-mark-reichsbanknote-1910-value/

Below is a 1920 10 *reichsbanknote,* the first note issued by the new Weimar Republic, which governed Germany from the abdication of Kaiser Wilhelm II in 1918 until Hitler consolidated his power and abolished the Republic in 1933.[13] Ralph labeled the note below as being worth as "$100." I'm not sure where Ralph got his exchange rate information, for by the hyperinflation era that peaked in 1923, it took over 4.2 trillion marks to equal one US dollar![14] It may be that Ralph was thinking of the exchange rate between marks and dollars under ordinary economic conditions.

13. https://collection.maas.museum/object/390060

14. https://www.johndclare.net/Weimar_hyperinflation.htm, quoted in https://en.wikipedia.org/wiki/Hyperinflation_in_the_Weimar_Republic

The inflationary impact on currency denominations is apparent in Ralph's 50 million reichsmark banknote (below) issued July 25, 1923, during the very height of the inflationary crisis during the Weimar Republic era. Paper money was being printed at such a furious rate that some bills were only printed on one side, such as the bill below — again mis-valued at "$5,000,000" — that Ralph brought home:

The 10 reichsmark bill below was printed from August 30, 1924, until January 22, 1929, after the worst of the inflationary pressures had eased. Pictured on the front of the bill is a portrait of Albrecht Daniel Thaer, a late 18th and early 19th century German agronomist — a specialist in soil management and crop production. With all due respect to the importance of good crops, Thaer must have been a really good agronomist to get his picture on national currency — or perhaps the government had printed so many bills that they were scraping the bottom of the barrel for currency portraits!

The bills on this and the next page are *rentenmarks,* issued after the end of the Weimar Republic when the Nazis had erased the last traces of German democracy in 1933. The new Nazi finance minister, Hans Luther, devised a way to place this new paper money on a solid footing as a medium of exchange by backing it not with scarce precious metals, but with real goods and property, including business and agricultural lands.

Ralph taped these souvenir rentenmarks into his album in a way that allowed the reader to see both front and back, revealing currency images that reflect popular Nazi imagery: a youthful, virile German male face that looks quintessentially "Aryan"; women working in the fields, indicating the importance of agriculture; a male laborer, which symbolizes heavy industry; and finally, the Brunswick Cathedral and Lion statue, both of which bring to mind German's history as an ancient, even mythical nation. The top is the front side of the bill, and the bottom is the reverse side.

Above: 1942 Five Reichsbanknote — Front

Above: 1942 Five Reichsbanknote — Reverse

Allied Military Currency (AMC)

This half-mark bill (above) that Ralph describes as "German invasion money" was in fact just that — money printed and issued by the Allies in occupied Germany. Allied authorities paid GI's with this money to avoid exacerbating inflation with US dollars that were much stronger than local currency.[15] The inscription at the top of the bill on the left, *Alliierte Militarbehorde,* translates as "Allied Military Authority." Called "AM-marks" or AMC (Allied Military Currency) for short, many of the bills were printed in Boston by the Forbes Lithographic Manufacturing Company.

Great secrecy shrouded the printing of the currency, to the extent of assigning the production process code names — in Germany, the code name was "Operation Wild Dog." AMC issued in other countries — Italy, Austria, Japan — each had its own unique code designation as well.[16] Ralph collected similar invasion notes in France and Luxembourg.

15. Rundell, Walter. 1961. "Currency Control by the United States Army in World War II: Foundation for Failure," Pacific Historical Review 30:4, (pp 381-399), as cited in https://azmemory.azlibrary.gov/digital/collection/msiartifact/id/14/.

16. https://www.educationalcoin.com/media/amfile/files/(1)imageshistoryfoliosalliedmiltary8bnfolio.pdf

Weimar Republic and Nazi Era Stamps

Ralph, as he himself declared, had neither "the patience nor the inclination to collect stamps." Likely his distaste for this arcane hobby explains his uncharacteristic failure to offer any explanatory background on any of the stamps that, as he delicately put it, he "picked up … from someone who USED to collect stamps over in Germany."

It is remarkable that Ralph successfully transferred all these stamps from their original location first into his 1951 album, and then from that deteriorated volume into his 1972 books, without damaging them beyond saving. Only a meticulous, detail-oriented (read OCD) fellow like my late, sainted father could have successfully completed what had to be a numbingly tedious chore! Here follows a sampling from the multiple pages of stamps Ralph brought home.

Nazi- Era Political Stamps

The allegorical stamp on the left below depicts a child being embraced in the arms of its mother and is captioned, *"Die Saar kehrt heim"* ("The Saar comes home"). It was issued on January 16, 1935, to celebrate the Saar Plebiscite that reunited that coal-rich region with the Reich. The Saar had been stripped from Germany and given to France by the Treaty of Versailles, one of the many humiliations imposed on Germany by the victorious Allies that Hitler had vowed to rectify.[17] There were many such commemorative Saar stamp issues, including the one on the right below featuring the word *"Saar"* clasped in hands representing the German nation.

---

17. https://www.stamp-collecting-world.com/thirdreich_1935.html

Paul von Hindenburg was one of the great heroes of the German people. General of the Imperial Army in WWI, Hindenburg was elected German president in 1925 and served until 1934. He tainted his legacy by allowing himself to be pressured to appoint Hitler Chancellor of Germany in 1933. The so-called Hindenburg Medallion stamps, like the one issued in January of 1933 (**preceding page**) were part of the Nazi effort to "be viewed in support of the ideals of patriotism, continuance and harmony during the transition from the democratic [Weimar] Republic to" the Nazi dictatorship.[18] It took nearly nine years for Hitler to feel secure enough in office to replace Hindenburg's image with his own, starting in 1941.[19]

The stamp pictured above, issued on July 25, 1935, celebrates a national meeting of the Hitler Jugend — something like a National Boy Scout Jamboree. Note the runic "S" on the banner beneath the trumpet. The caption at the top, *"Wettreffen der H.J.,"* translates as, "Meeting of the Hitler Youth."[20]

---

18. https://www.rfrajola.com/PDR2017/Coffey1/Coffey1.pdf

19. https://www.stamp-collecting-world.com/thirdreich_hitlerdef.html

20. http://www.philatelicdatabase.com/hitler/stamps-of-germany-world-jamboree-of-hitler-youth-issues-1935/

The 1936 Olympics afforded Hitler a propaganda stage on which to display the supposed superiority of the Aryan Race and the Third Reich. The Regime made sure the world got its invitations via stamps such as this one issued in November of 1935 (**above left**).[21] **Above, right**: On January 26, 1943, the Nazis celebrated the tenth anniversary of their rise to power with this stamp, which bears an image of the Brandenburg Gate surmounted by the Nazi eagle grasping a swastika in its talons. [22]

---

21. https://www.stamp-collecting-world.com/thirdreich_olympics.html

22. https://www.stamp-collecting-world.com/thirdreich_1943.html

Nazi-Era Commemorative Stamps

Pictured above are stamps honoring, from left to right: (1) auto magnates Gottlieb Daimler and Carl Benz (issued in 1936 to celebrate the German Auto Show in Berlin); (2) Frederick the Great, a much-admired Prussian emperor (reign 1740 – 1772) whose legacy Hitler tried to co-opt in order to bolster the legitimacy of his own Reich;[23] and (3) writer, scientist and statesman Johann Goethe (1749 — 1832); the Nazis loved to celebrate the "superior" nature of German culture by promoting works by playwrights such as Goethe.[24]

***Wintershilfswerk* Stamps:** The stamps above commemorate an annual charitable drive initiated on behalf of Germans suffering during the Great Depression under the Weimar Republic and continued by

23. https://en.wikipedia.org/wiki/Frederick_the_Great

24. https://encyclopedia.ushmm.org/content/en/article/culture-in-the-third-reich-disseminating-the-nazi-worldview

the Nazis under the auspices of the National Socialist People's Welfare. Donations to this ostensibly private welfare agency were supposedly voluntary, but were in practice mandatory; in this way, more of the public tax money could be funneled into the Nazi war machine. *Wintershilfswerk* translates as "Winter Relief" — the full name was *Wintershilfswerk des Deutschen Volkes* ("Winter Relief of the German People"), or, as it was more commonly referred to, *WHS*.

Nothlife (above) **Commemorative Stamp:** This is one of the first sets of commemorative stamps issued by the Nazis toward the end of 1933. *Nothlife* — inscribed vertically along the right edge of the stamp — means "emergency assistance." A portion of the sale price for the stamps went to various charitable causes. 1933 also happened to be the tenth anniversary of the first *nothlife*.[25] This stamp celebrates Richard Wagner's opera *Die Meistersinger*. Hitler was a great devotee of Wagnerian opera, believing that its Germanic mythological motifs supported his doctrines of Aryan superiority.

The stamps **on the following page** each bear at the top the caption *Kolonial gedenk jahr* — "colonial commemorative year." The Nazis issued these stamps in 1934 to mark the 50th anniversary of the beginning of the German colonial empire. The three-pfennig stamp on the left displays the image of Franz Adolph Luderitz (1834 — 1886), who established the German colony of German Southwest Africa. The twelve pfennig stamp bears the likeness of Karl Peters, co-founder of German East Africa.[26]

---

25. https://www.stamp-collecting-world.com/thirdreich_nothilfe.html

26. https://www.stamp-collecting-world.com/thirdreich_1934.html

Weimar Republic-Era Stamps

The "Wooden Dove" (**above**) is a 1924 100-mark German airmail ("*deutsche flugpost*") stamp. This particular stamp fetches anywhere from $19 to $119 on the collectors' market, depending on the web site and the condition of the stamp.[27]

---

27. https://www.hipstamp.com/listing/70457-germany-reich-1924-airmail-wooden-dove-birds-mi344-350-mint-re-gummed/19552485

The Ravages of Inflation during the Weimar Era

By mid-1923, inflation during the Weimar Republic was at its worst, with catastrophic effects on ordinary citizens, as the price of all consumer products, including stamps, went into the stratosphere. The cost of mailing a letter reached as high as two million marks, an amount unimaginable to the modern mind. Inflation raged so out of control that it outpaced the printing presses — stamps became worthless as soon as they hit the market. So, instead of reprinting new stamps, the government simply over-wrote existing stamps with the new revaluation. Above are four such stamps, revaluated from thirty marks to 20,000 on the left, and from forty marks to 5,000 on the on the right.[28]

The stamps on the **next page,** also from the hyper-inflation era, are for 300, 400, 4,000 and 5,000 marks. It's easy to see why the authorities finally gave up trying to keep pace with the economic crisis, resorting to over-writing with the new inflated prices.

As an aside, the term *Reich* was not unique to the Hitlerian era. The word simply translates, "empire," or "nation," and was used by both the Imperial and Weimar governments. Hitler's use of the expression "Third Reich" refers to his claim that his new Nazi order was the successor to what he saw as the two great predecessor empires in German history, the Holy Roman Empire (A.D. 962 — 1806) and the Imperial German Empire that fell after WWI.[29]

---

28. https://www.stamp-collecting-world.com/weimarrepublic_hyperb.html

29. https://en.wikipedia.org/wiki/Nazi_Germany

Witnesses to History

Ralph is, of course correct — this relic (**above**) is indeed a German police stamp. While Ralph provided no further details about where, how or from whom he obtained this item, a bit of detective work has revealed a possible solution to this mystery.

The *ordnungspolizei* — *Orpo*, for short — was the uniformed law enforcement agency under the Nazi regime. This nationalized police force had its headquarters in the Berlin offices of the much-feared SS, headed by Hitler's infamous henchman, Heinrich Himmler.

Under the swastika on the stamp are the words, *Chef der Ordnungspolizei* — "Chief of the Order Police." The words around the circle tell us more about that chief: *"Der Reichsfuhrer SS und Chef der Deutschen Polizei* — "Reich Leader and Chief of the German Police" — a title belonging to none other than Himmler himself.

Himmler's family home was in Gmund am Tegernsee in Bavaria. As it happens, Ralph was in Finsterwald on May 2, 1945 — a village located on the northwestern edge of Gmund am Tegernsee! Is it possible that Ralph, unaware at the time of whose house lay only a stone's throw east of his half-track, picked up an official document stamped by Himmler himself off the ground or from an evacuated local *Orpo* office?

Above left: Heinrich Himmler **Above right:** Himmler's house in Gmund am Tegernsee

Above on the right is a page taken from a business personnel register book. *Arbeitsbuch* means "workbook," so Ralph got that part right. I once thought this might be a government record book of some kind due to the swastika watermark, but it is not. World War II historian Albert Trostorf's uncle, Franz Josef Trostorf, had a similar book (**above, left**).

Vater, wenn es möglich ist, laß diesen
Kelch an mir vorüber gehen. doch nicht
mein, sondern Dein Wille geschehe!

Luc. 22. 42

Mit Genehmigung der Photographischen Gesellschaft Berlin·Ch.

RELIGIOUS CARD

Above: A devotional card Ralph brought home. Despite the quasi-pagan, anti-Christian ideology of the Nazis, many ordinary Germans were quite religious — predominantly either Catholic or Lutheran, depending on the region of the country in which they lived. This card quotes from Luke 22:42 — "Father, if thou be willing, remove this cup from me; nevertheless, not my will, but thine, be done." The caption at the bottom reads, *"With permission of Gessellschaft Photographers, Berlin"*

Epilogue: Building a Nation

"They married in record numbers and gave birth to another distinctive generation, the Baby Boomers. They stayed true to their values of personal responsibility, duty, honor, and faith..."
— **Tom Brokaw**

Their duty done and done well, the Greatest Generation returned home from war and immediately set about building the Greatest Nation. They did so with the same dogged determination, indomitable spirit and humble attitude that had carried them to victory in war. Some of them, as when they wore Army green, did spectacular things once they donned civilian clothes. But most did countless ordinary things — got jobs, worked hard, loved their families, and generally went about their business as loyal and productive citizens of the country they had defended at the risk of their lives.

On the following pages we take a glimpse at the pre- and post-war stories of some of these humble heroes — men who had jounced along in a radio half track across the muddy fields of Europe, dodging bullets and bombs, faithfully doing their "little" part in the Big Picture of what has been termed "the Good War."

Ralph knew each of these men intimately, trusted them totally, and formed with them the kind of bond that can only be forged in the hot crucible of combat. Like Ralph, they now all sleep quietly in cemeteries from Washington to Georgia, from Michigan to Alabama.

But if we listen carefully, very carefully, we can still hear their message. We hear it in the mournful sound of the trumpet echoing out "Taps." We hear it in the sharp snapping of flags as they ripple in the stiff breezes that blow across the bivouacs of the dead in the Punch Bowl and in Henri Chapelle. We hear it in the shuffling gait of their few surviving comrades as they march — or ride, if they must — in parades in small towns and big cities each Memorial Day and Veterans Day.

We began this story with their message, and it seems altogether fitting and proper that we end with it:

Earn this!"

Half-track Radio Group, Radio Section, HQ and HQ Company, 12th Infantry Regiment, 4th Infantry Division — 1944-1945

Ralph recalled **T/4 Charles Manton Emery, Junior** as "a quiet and efficient young fellow from Erie, Pennsylvania." And so he was. Emery was born on Christmas Eve, 1921, to Charles Senior and Janet Emery — making him just a few months younger than Ralph, who was born in March of that year — in the quiet little burg of St. Mary's in Elk County, Pennsylvania. The family soon moved about seventy miles west to Erie, Pennsylvania, where they put down deep roots. Before entering the Armed Services on August 7, 1942, Emery worked as a metal fabricator, perhaps accounting for his "efficient" manor, a trait that would have served him well in his civilian job. After his discharge from the Army at Camp Butner, NC, Emery returned to his home for good in Erie, where he took a job as a clerk in a dry goods store, apparently remaining a life-long bachelor until his death in March of 1987.

T/Sergeant Raymond W. Kimmel, remembered Ralph, was the guy "who was forever running up with unpleasant news," for it was his task as Radio Section Chief to bring all the "glad" tidings from higher-ups to the half-trackers — usually orders to move out on-the-double to heaven-knows-where next. Good thing he was a good Joe, or the guys may have been tempted to shoot the messenger! Born on February 21, 1907, thirty-seven-year-old Kimmel was the old man of Ralph's group of mostly twenty-somethings. He came from sturdy blue-collar stock in Lebanon, Pennsylvania, laboring as a steelworker at Bethlehem Steel in that city. After his discharge, Kimmel left his steel-factory roots to work as an electrician at a number of different electrical companies — perhaps his time spent using and repairing walkie-talkie radios in Europe convinced him that tinkering with wires held more charm than wrestling with huge ladles of molten steel in a hot, noisy factory. T/Sgt Raymond Kimmel lived out his years in Lebanon, passing away on

May 3, 1974. His obituary lists no wife or children among his survivors, but it does make one very eloquent and powerful statement about the man: "He was a veteran of WW II." Kimmel rests beneath the green sod of Holy Cross Cemetery in Lebanon under a modest headstone inscribed, "T/Sgt, HQ and HQ Co, 12th INF, WW II."

Twenty-nine-year-old Ohioan **T/5 William Donald "Bill" Harris** was the driver for the squad, which meant he also had the responsibility of keeping the half-track in working order. Whenever there was a lull in the action, Ralph wrote, Harris would immediately jump out of the truck and assume "his typical position with his nose buried under the engine hood." Harris would then set about pulling things apart — and then frantically attempt to put it all back together before Kimmel could arrive with orders to move out. Army records show that Harris enlisted on March 8, 1941, as a single man. He married twice after the war, but available records only show his second marriage, to Edith McKinnon, whom Harris wed in 1990 at age 74. He, like many of his comrades, returned to his roots in Zanesville, Ohio, living all of his 94 years in that city and passing away on June 25, 2010. He is buried in Memorial Park Cemetery in Zanesville.

Jack R. Verga presents a bit of an enigmatic figure. He appears on Gerden Johnson's HQ and HQ Company roster, yet without any rank designation. However, his name appears nowhere in the 12th Regiment Yearbook published in 1946 — but the Yearbook, though a rich source of information, does have a number of gaps in its personnel listings. The good news is, Jack appears frequently in public ancestral records and censuses, which afford us a good feel for his background, life, and formative experiences.

Born on September 16, 1919 in The Bronx, New York City, to Frank and Caroline Verga, Jack was a second-generation American whose grandparents, Jack and Mary Ann, immigrated to the US from Italy. Sadly, Jack had an older brother, also named Jack, who was born and died in 1915 — apparently "our" Jack was named in his memory. Jack was a slight fellow, standing at a diminutive five feet, four inches and weighing a slender 125 pounds, according to his 1940 draft registration card. By the time he actually entered the service one year later at Ft. Jay, he had grown an inch and gained five pounds! Before Uncle Sam claimed his services, Jack worked in radio repair, service and sales for Boulevard Radio Sales and Service in The Bronx, and he was living a bachelor's life in Nassau, NY. With that experience and background, it is no wonder that he was, as Ralph observed, "an inveterate fixer." After the war, Jack lived a good many years in Garden City in Nassau County, New York, possibly until the year 2000. Public records indicate that at that same address, a "Jack R. Verga" was born March 20, 1973 — assuming that this is Jack's son, Jack would have been 54 at the time. There are no more City Directory entries for Jack after the year 2000, which suggests that Jack passed away during or before that year while in his late seventies or early eighties.

Ralph spoke often and fondly over the years of **T/4 Robert Taylor "Bob" Weltzein,** whom he considered "my very best friend in the Army." When Ralph and his family moved to Portland, Oregon, in 1960, he made a point to travel to Seattle and spend some nostalgic time with his erstwhile tentmate. It comes as no surprise that, out of all the guys in his unit, Ralph formed a particular bond with Weltzein, the only college graduate in their group. In temperament and in outlook, they were very much kindred spirits — both quiet, serious, cerebral, and bookish, with similar ideas about right and wrong and life's core values. On the lighter side, Ralph also described Weltzein as a fastidious kind of guy, being "most meticulous about his personal appearance."

Weltzein was born in Seattle, Washington on May 1, 1921 — making him almost exactly Ralph's age. He entered the Army in April of 1943, served with the 12th Infantry in Europe, and left the service a couple of months after Ralph, on December 12, 1945. Following his discharge, Weltzein returned to his native Northwest, settling in Seattle with his bride, Lorrane Cushing Boos, whom he married in 1946, six months after coming home from the war. T/4 Robert T. Weltzein passed away in Washington on July 19, 1997.

Born on February 13, 1920, to Italian immigrant parents, Key Stone State native **T/Sergeant Nunzio Yocca** was the leader of Ralph's half-track radio group, and he impressed Ralph as "a very fine young man." Before being drafted in January of 1942, coal-miner's son Yocca worked as a semi-skilled mechanic and repairman, which explains Ralph's description of him as "a very good handy man." Yocca was one of the few in Ralph's group that had not arrived in Europe as an Infantry Replacement but had landed with the Fourth Infantry Division on D-Day at Utah Beach. Yocca, after his discharge from the military in July of 1945, obtained a teaching degree in industrial arts — popularly known as "shop class."

Yocca had a very distinguished post-war career as an educator at Windber Area High School in Windber, Pennsylvania, retiring after thirty years of service. Yocca was very active in his church and in the community, and he was a member of both the American Legion and the VFW. In recognition of his many contributions to his school and his town, Yocca was inducted into the Windber Hall of Fame in 2009. He married Floria Dorina Santucci in 1952, and they lived until their deaths (she in 2002 and he in 2010) in Windber, where they are buried in Saint Anthony's Cemetery. Yocca's obituary notes that T/Sgt Nunzio Yocca was the "winner of numerous medals and awards during his military career." Included here is a picture of Yocca in the faculty section of the 1974 Windber Area High School Yearbook.

M/Sergeant Woodrow Wilson Bledsoe

"... Sgt Woodrow Bledsoe of Altoonia [sic], Georgia, was the first to try to rescue us, but he got a direct mortar hit on his jeep. I yet do not know whether or not he made it."
— Ralph Miles, 1951.

"Courage is when you're scared to death, but you saddle up anyway."
— John Wayne, Actor

Using John Wayne's definition of bravery — as eloquent and accurate as any — M/Sergeant Woodrow Wilson Bledsoe was indeed a very courageous man. Bledsoe, at great risk to himself, "saddled up" his jeep and drove straight into a heavy German artillery and mortar barrage during the Battle of the Hurtgen Forest in November of 1944 to rescue Ralph and three of his comrades from being blasted to atoms in an exposed forward position.

Thankfully, Sgt. Bledsoe did in fact survive the explosion, but his wounds were serious. The mortar round that landed practically in his lap miraculously did not take his life, but it did shatter his left femur, radius, and ulna, requiring extensive treatment and months-long convalescence in a hospital in Europe. By March 19, 1945, Bledsoe was well enough to be placed on board the SS *Queen Elizabeth* in Gourock, Scotland, and sent stateside, where he received an honorable discharge and a number of well-deserved military honors for his actions.

Following his recovery from his wounds, Bledsoe married Helen Pappas and settled in Stone Mountain, Georgia, where they raised their daughter, Helen Bledsoe Adkins. Bledsoe passed away at age 89 on May 14, 2008. He rests in his hometown of Georgetown, Georgia, beneath a military marker that reminds his descendants of the price he — and tens of thousands of others — paid for our freedom.

Combat veterans love their country and its ideals, but what they really fight, bleed and, if need be, die for, is each other. So it was with Woodrow Wilson Bledsoe. Ralph would have been profoundly relieved to discover that his fellow Georgian did not pay with his own life trying to save his. So am I.

Here's a slow hand-salute to a personal hero whom I never met. RIP, soldier.

Following his Army discharge, **Ralph Miles** returned to his job as night clerk at the Savannah FBI. He was soon promoted to day clerk and then to Chief Clerk, a position of great responsibility in the Bureau, involving supervision of all clerical staff and maintenance of all office records and communications.

No stranger to grinding hard work, Ralph doggedly juggled his duties with the Bureau (usually putting in sixteen-to-eighteen-hour days), studying law into the wee hours of the night and teaching at church. Much to Eleanor's exasperation, Ralph refused to use the GI Bill to help with expenses, disdaining the money as a "government hand-out." *"The only helping hand I need or want is the one at the end of my own arm,"* he often vowed.

Appointed a Special Agent of the FBI in 1951, Ralph served with distinction at various offices across the nation, rising rapidly through the ranks. He ultimately was assigned to the FBI office in Birmingham, Alabama, as Special Agent in Charge of the northern half of the state in 1968. While serving in Birmingham, Ralph hosted the two FBI Directors appointed by President Nixon following the death of J. Edgar Hoover, a man Ralph greatly admired and respected: L. Patrick Gray, who resigned in the wake of the

Watergate scandal that engulfed the Nixon Administration, and then Clarence Kelley, a good man and a former FBI man himself.

Above left (left to right): Eleanor, FBI Director L. Patrick Gray, Mrs. Gray, Ralph. **Above foreground, L-R:** Eleanor, Ralph, FBI Director Clarence Kelley

Ralph retired from the Bureau in 1975 following two serious heart attacks, settling in Birmingham. Ralph passed away on March 10, 2008, just eighteen days shy of his 87th birthday. He rests beneath a military footstone in Southern Heritage Cemetery in suburban Birmingham.

Appendix

Reproduced here and on the following pages are Ralph's diary notes recording events from his induction in 1943 through his return to the States in June of 1945. The diary contains considerable detail as to dates, places and times — even the weather conditions during the launch of Operation Queen on November 16, 1944, in the Hurtgen Forrest. After that, Ralph's notes are less detailed but still insightful, especially when taken together with his other recollections (both oral and written), as well as the many relics he included in his memoirs.

Doubtless his early notes were more involved because infantry replacements had a lot of time on their hands to while away. Of course, that all changed once Ralph got his assignment with the Division and Regiment and entered into the belly of the combat beast.

Above: Ralph scribbled notes on whatever scraps he could find, with the result that the sheets don't always follow chronological order

Property of Ralph J. Miles, 34828227.
In case of -- please notify Mrs. R.J. Miles, 516 West 41st
Street, Savannah, GA, U.S.A. - Faith Church of Christ - P.
.$10,000 ins., Allotments: (L.F. $22, Cl.E. $21 to wife - 2 children.

9-8-43 = Inducted, Ft. McPherson, Atlanta, GA, returned home for 21 days.
10-4-43 to 2-10-44 (appx. At Ft. McClellan, Ala. - Basic + Specialist, G15 5
2-11-44 to July 11-44 (appx. Ft. Benning, GA. - ECC 42, 230 Co., 1st Str and
    ERRC 7, 22nd Co., 1st Str
7-24-44 to 8-1-44 = Ft. Geo. G. Meade, Md. Repl. Depot., Co. B, 1st Bn. St Rgt.
8-2-44 to 8-6-44 = Camp Shanks, N.Y., Co. D (4th Pl.), shipment GN 9 Co a.
8-7-44 to 8-14 = Lv. N.Y. via HMS Aquitania, 8,000, and Ar. Greenock, Scotld.
8-14-Mon. = Lv. Greenock via train and Ar. Wellington, Eng. 8-15-Tues.
8-17-Thrs. = Lv. Wellington via GI truck + Ar. Cmp Stapley 8-17- Co. EE, Pkg. X-70-G
    (204th Co. in Wellington).
8-20-Sun = Lv. Stapley via train + Ar. Southampton 8-20
8-20-Sun = Boarded HMS Devonshire + Lv. Station Mon-8-21+ Ar. Coast of France
    same day.
8-22-Tues = Landed Omaha Beach near Trevieres by LCT.
8-25-Fri = Lv. Trevieres (182.6, 42 Bn) via truck + Ar. 8. Mi. LeMans, 485 Co., RT 6 N S
9-1-Fri = Lv. LeMans via truck noon Ar. Mortagne dusk, 2346, 90 Bn, 50 R.D., APO 153
9-8-Fri = Lv. Mort. once via truck, Ar. Melun 0900,   "   "   "   "
    Saw Bing Crosby show. 600 Seat.
9-9-Sat. = Lv. Melun 0015 +   "   "   Soissons 0900,   "   "   "   "
9-10-Sun. = Saw waves bombers all morning & toured line. Heard expl. all day.
9-11-Mon. = Took TMR. hike saw badly bombed Barisis and RR worse bombed.
    Living with Joe "Doc" Behan, Bill Maddox, Don Adams, "Boots" Biziane,
    Walter J. Blaze.
9-15-Fri = Behan + Adams shipped. We Lv. Soissons 9pm. Convoy stopped by
    4 Jerries (according to Frenchman).
9-16-Sat = Ar. 20mi. W. of Liege, Belgium. Worked KP. Saw explosion 1000 yds away.
9-17-Mon = A convoy strafed 20 miles away.
9-18-Tu. = Saw for "Going My Way", interrupted by enemy planes. Bombs + flak
    close. Saw hundreds of bombers all morning going toward
    line + returning. Thousands go over almost daily.
10-6-Fri = Air activity extremely hvy. Flare dropped 100 yds. Steady thunder
    of art. + bombs became little more distinct. Last few days had
    heard little though they were constant + close first 2 wks.
10-17-Tues = Lv. ABEE noon Ar. 41st Bn, 310 Co, 3rd RD, 3:30pm, near Malmedy in
    Ardennes Mts. 35 mi S of Aachen + 5 mi in Belg.
10-18-Wd. = Lv. Co. 310 at 5:30pm Ar. Ser. Co., 4th Div, 12th Inf. 6pm; Lost John Nivala.
10-19-Thr = Lv. Ser. Co. + Ar. Hq. Hqs. Co., 12th Inf 12 noon. Assigned Rad. Com. Saw Glose + new
11-6-Mon = Lv. 6pm Ar. SL 1200 11-7-Tues. Sleeted. Heurtgen Forrest.
11-8-Wed = Rained + snowed all night + day of 11-9.
11-16-Thr = Atk launched 1100. Cl. day.
12-8 Fri = Lv. Germany Ar. Luxembourg dusk. Junglunster
12-27 Mo Lv. Jung ... Junglinster
12-28 Th Lv. Sand Ar. ...

Above: Ralph made sure he remembered the name of the GI that he feared had given his life while attempting to bring Ralph and others back from the front in the Hurtgen (see Chapter 5)

Above: Ralph's scribbled notes of the assault on Prum, R&R in Verrerie, crossing the Rhine at Worms, Task Force Rodwell ("TF"), the drive into Bavaria ("on the move ... "), and ending with his post-war movements to Windsheim, Bamberg and Le Havre.

NO HERO
ONLY EUROPE - INFANTRY - SQUAD TO DIVISION
9-43 DRAFTED - 22½. MARRIED, 1 CHILD 3 MO.
           BASIC - FT. McCLELLAN + RADIO
           FT. BENNING, GA. ADVANCED RADIO + REPAIR
                 6-6-44 INVASION - ME AT FT. BENNING
   8-44 ACQUITANIA TO SCOTLAND - 8,000 MEN. ZIG ZAG
       •   FRANCE. UTAH BEACH. PILES EQUIPMENT. RUSTING HULKS
AUG - OCT.   REPLACEMENT CAMPS.   BUZZ BOMBS. OUR PLANES
10-19-44   12 RGT, 4TH INF DIV. - HOLZHEIM, GERMANY
11-44   HURTGEN FOREST NR AACHEN, GERMANY. COLOGNE
       "VOLUNTEERED" TO LINE. T FORWARD OBSERVER
       LIVED IN HOLES - LOGS -
           FIRST-HAND CASUALTY

Left: Index cards Ralph used when speaking to school groups about his war experiences.

Notice his opening line: "*No hero.*" This was a common feeling among GI's who survived the war. The true heroes, they all firmly believed, were those who *didn't* come home.

2

HURTGEN FOREST .
       TREE BURSTS   BOOBY TRAPS   PNEUMONIA   FROZEN FEET/HANDS.
12-8-44 TO LUXEMBURG. DEFEND RADIO STA.
           PASS INTO LUX CITY
12-16-44 BATTLE OF BULGE
       BOOK — COMBAT INF BADGE
JAN 45TH. REVERSED BULGE.   BASTAGNE.   PATTON
3-1-45   PRUM   FROZEN GERMN SOLDIERS
3-45   REST IN FRANCE - A WEEK.
MAR-APRIL   CENTRAL GERMANY, THEN SOUTH TOWARD BAVARIA
           AIRPLANES ALONGSIDE AUTOBAHNS
       WASSERAL FINGEN — BOOK
   HITLER YOUTHS
   POLISH SLAVE LABORERS

Left: "Polish slave laborers" (arrow) likely refers to the refugees rescued jointly by the 522nd Field Artillery Battalion and the 3rd Battalion of the 12th Regiment on May 2, 1945.

3

ETO WAR ENDED 5-8-45.

WEEK PASS TO RIVERIA, FRANCE

JULY —   SENTIMENTAL JOURNEY

HIROSHIMA 8-6-45   JAPAN SURR 8-14-45

10-14-45 DISCHARGED AS STAFF SGT

BOOK - #2 - BACK PAGES

Above: Ralph's discharge documents: Ralph was actually 6' 1" tall — Army didn't see it that way, somehow!

The Army used the Adjusted Service Rating (ASR) score during the post-war demobilization effort. Also called the "point system," it returned troops back to the U.S. based on the length of time served, family status and honors received in battle. Ralph had a score of 78. (https://work.chron.com/army-asr-score-22134.html)

MILITARY EDUCATION

14. NAME OR TYPE OF SCHOOL—COURSE OR CURRICULUM—DURATION—DESCRIPTION

Attended the infantry school, Ft. Benning, Georgia for 20 weeks. For
14 weeks was enrolled in Radio Operator course, receiving a superior
rating. Studied radio theory, installation, and operation of infantry
radio sets.
      For six weeks enrolled in radio repairman's course. Studied radio
theory, repair and maintenance of infantry radio sets. Received an
excellent rating.

CIVILIAN EDUCATION

15. HIGHEST GRADE COMPLETED	16. DEGREES OR DIPLOMAS	17. YEAR LEFT SCHOOL	OTHER TRAINING OR SCHOOLING	
			20. COURSE—NAME AND ADDRESS OF SCHOOL—DATE	21. DURATION
12th Grade	H.S. Diploma	1940	NONE	NONE

18. NAME AND ADDRESS OF LAST SCHOOL ATTENDED
Commercial High School
Savannah, Georgia

19. MAJOR COURSES OF STUDY
      Commercial

CIVILIAN OCCUPATIONS

22. TITLE—NAME AND ADDRESS OF EMPLOYER—INCLUSIVE DATES—DESCRIPTION

CHIEF CLERK: Supervised and coordinated the clerical work of a field
office of the F.B.I. Supervised the keeping of all records, filing,
issuance of supplies, requisitioning of equipment, taking and keeping
      inventory records and coordinated work of 6 clerks. Worked for
Federal Bureau of Investigation, Savannah, Georgia.

NOTICE OF CLASSIFICATION
Ralph   Jacob   Miles
Order No. 10972 has been classified in Class 1-C Disc.
(Until   19   )
by X Local Board
   Board of Appeal (by vote of
   President
10-22-45, 19

ADDITIONAL INF

23. REMARKS
                     NONE

24. SIGNATURE OF PERSON BEING SEPARATED   25. SIGNATURE OF SEPARATION CLASSIFICATION OFFICER   26. NAME OF OFFICER
Ralph J. Miles   B. L. B.   H.L. BONNIWELL, 2nd Lt. MAC
                                                GEN SEPARATION POINT

Above: Second page of Ralph's discharge papers. Attached to the document is the most coveted piece of paper of all: "Notice of Classification: 1-C Discharged"!

Many years after the war, Ralph arranged all his military medals and other Army memorabilia onto a matte, labelled each item and then framed the collage. Those items can be seen on this and the following page. The captions, except as otherwise noted, are Ralph's own.

Above: Patch of the Fourth Infantry Division

Below: As an infantry soldier in World War II, the below insignia were worn on tunic or shirt collars

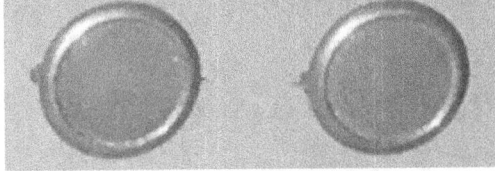

Above: My final year at Savannah High school, I was the ROTC Adjutant, a 1st Lieutenant, and wore these buttons on tunic shoulders and blouse collar.

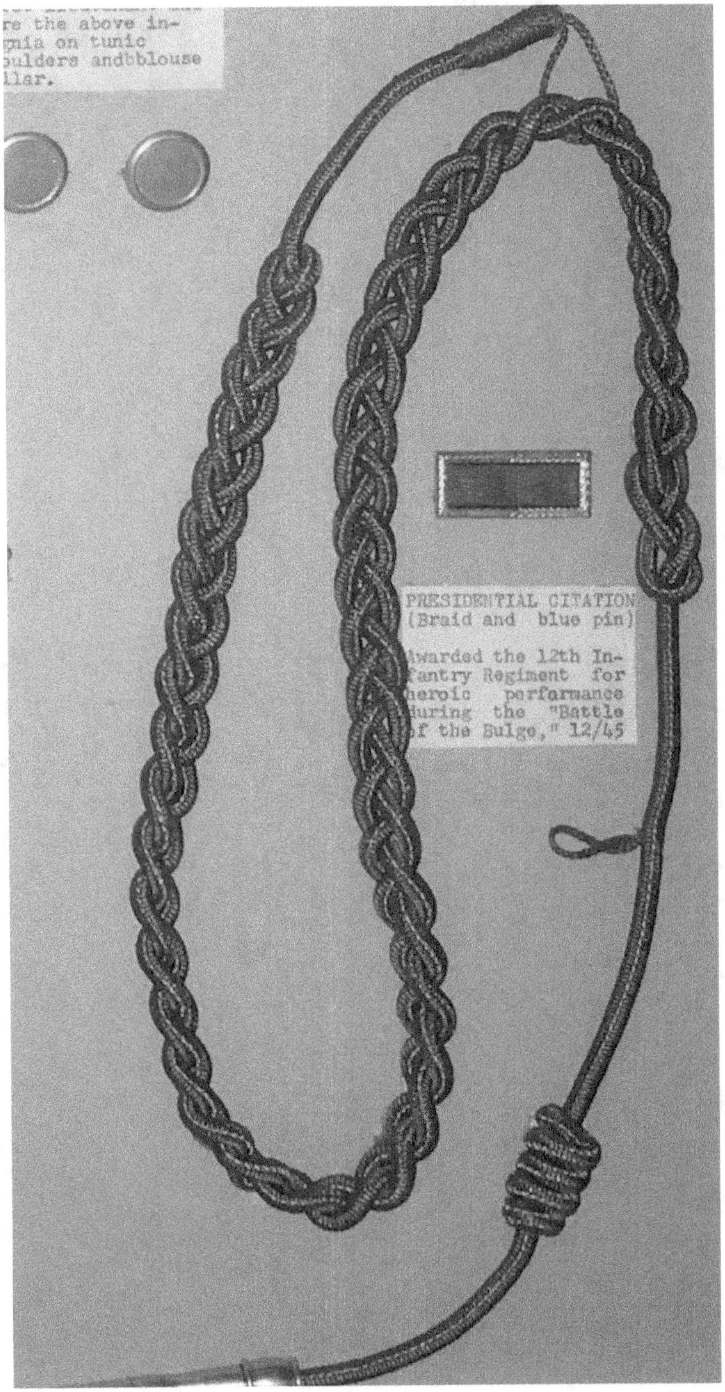

Right: Presidential citation (braid and blue pin). Awarded the 12th Regiment for heroic performance during the "Battle of the Bulge," 12/45.

World War II Victory medal

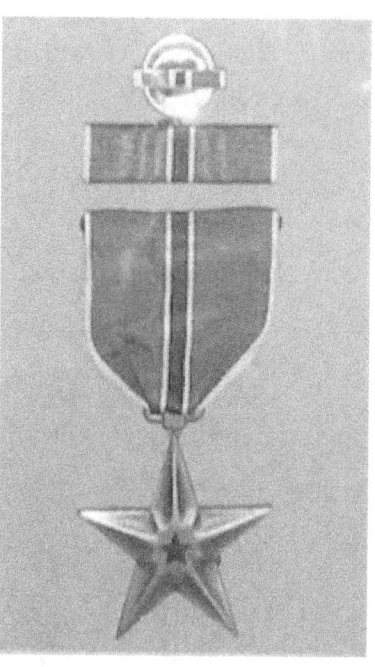

World War II Service in the United States medal.

World War II Army of Occupation in Germany medal

World War II Service in the European Theater of Operations medal

Bronze Star. Awarded to all holders of the Combat Infantryman's Badge.

Military Honors Earned by Ralph J. Miles

ROTC – Savannah Commercial High School, 1937 - 1939

1. ROTC Silver Medal
2. ROTC Adjutant, 1st Lieutenant
3. Military Leadership Award

World War II — European Theatre of Operations
September 7, 1943 – October 14, 1945

1. World War II Service in the European Theatre Medal with Oak Leaf Clusters (for participation in four European Campaigns)
2. World War II Service in the United States Medal
3. World War II Victory Medal
4. World War II Army of Occupation in Germany Medal
5. Bronze Star
6. Combat Infantryman's Badge
7. Good Conduct Medal
8. Presidential Unit Citation Pin and Belgian Fourragere Braid
9. Promoted to Private, First Class
10. Promoted to Staff Sergeant
11. Honorable Discharge from the U.S. Army

Private First Class Ralph J. Miles, Senior, United States Army, age 23 — One of the Greatest Generation. Steadfast and Loyal. Led by Love of Country.

Bibliography

"About: Operation Undertone." *DBpedia,* https://dbpedia.org/page/Operation_Undertone

"After Action Report — 11th Armored Division, March 1-31, 1945." *The 11th Armored Division Legacy Group,* http://www.11tharmoreddivision.com/history/march_after_action.htm

"Alexander Patch." *Wikipedia*, https://en.wikipedia.org/wiki/Alexander_Patch

"American Logistics in the Normandy Campaign." *Wikipedia,* https://en.wikipedia.org/wiki/American_logistics_in_the_Normandy_campaign

An Illustrated History of Fort George Gordon Meade. (Fort George G. Meade, MD: The Fort Meade Museum, 1985).

Atkinson, Rick. "The Hurtgen Forest: The Worst Place of Any." *HistoryNet,* https://www.historynet.com/the-hurtgen-forest-1944-the-worst-place-of-any/

Babcock, Bob. *Action Against the Enemy, Reports After/After Action Reports for the month Of December, 1944, dated 6 January 1945*, at FB.com, National Fourth Infantry Division Association, December 16, 2020.

"Battle of the Hurtgen Forest." *Wikipedia*, https://en.wikipedia.org/wiki/Battle_of_Hürtgen_Forest

Beranty, Richard. "K Rations Created the World's Best Fed Army." History.com, June, 2013. Warfarehistorynetwork.com, June, 2013, https://warfarehistorynetwork.com/article/k-rations-created-the-worlds-best-fed-army/#:~:text=The%20concept%20behind%20K%20Rations,for%20and%20packaging%20K%20Rations.

Butler, Stephanie. "How Hershey's Chocolate Helped Power Allied Troops in WWII." History.com, March 28, 2023. https://www.history.com/news/hersheys-chocolate-allied-d-day-rations-wwii#

Brown, Jessica Wambach. "Camp Shanks: Last Stop USA." *HistoryNet*, October 23, 2020, https://www.historynet.com/camp-shanks-last-stop-u-s-a/

"Camp Shanks." *Wikipedia*. https://en.wikipedia.org/wiki/Camp_Shanks).

"Casual (US DoD Definition)." *Military Factory*, https://www.militaryfactory.com/dictionary/military-terms-defined.php?term_id=857

Certain, Geni. "Fort McClellan." *Encyclopedia of Alabama*, http://encyclopediaofalabama.org/article/h-3525.

Chapman, Craig S. *Battle Hardened: An Infantry Officers' Harrowing Journey from D-Day to VE Day*. Washington, D.C: Regnery History, 2017.

Chen, C. Peter. "Gerd von Rundstedt." *World War II Database*, https://ww2db.com/person_bio.php?person_id=29

Chen, C. Peter. "The Battle of the Hurtgen Forest 19 Sep 1944 — Feb 1945." *WWII Database*, https://ww2db.com/battle_spec.php?battle_id=117.

"Cigarette Camps: Introduction." *Skylighters*, http://www.skylighters.org/special/cigcamps/cigintro.html

Clare, John D. "Hyperinflation." *Johndclare.net*, https://www.johndclare.net/Weimar_hyperinflation.htm, quoted in "Hyperinflation in the

"Combat Infantry Badge CIB." *HRC: 'Soldiers First!'*, July 7, 2022 https://www.hrc.army.mil/content/Combat%20Infantryman%20Badge%20CIB

"Condolence Letters — Lieutenant John W. Irvine, 4th Infantry Divisionz91944)." *Omnigatherum — Antique Documents,* http://omnigatherum.ca/wp/?cat=35

"Culture in the Third Reich: Disseminating the Nazi Worldview." *Holocaust Encyclopedia,* https://encyclopedia.ushmm.org/content/en/article/culture-in-the-third-reich-disseminating-the-nazi-worldview

"Desertion." *Wikipedia,* https://en.wikipedia.org/wiki/Desertion

"Death Marches During the Holocaust." *Wikipedia*, https://en.wikipedia.org/wiki/Death_marches_during_the_Holocaust

"Death Marches." *Holocaust Encyclopedia*, https://encyclopedia.ushmm.org/content/en/article/death-marches.

Deutsche flugpost." *Hipstamp*, "https://www.hipstamp.com/listing/70457-germany-reich-1924-air-mail-wooden-dove-birds-mi344-350-mint-re-gummed/19552485

"Deutsche Wehrmacht Armband — Issue Stamp." *Epic Artifacts*, https://epicartifacts.com/product/deutsche-wehrmacht-armband-2/

"Evacuations of children in Germany during World War II." *Wikipedia*, https://en.wikipedia.org/wiki/Evacuations_of_children_in_Germany_during_World_War_II

"522nd Field Artillery Battalion." *Ibiblio*, http://www.ibiblio.org/45wwiiresources/522/522.html?fbclid=IwAR2IlEG6ZaMm5zQJZWBDgdc9OO7DTt-FYbLGaTkQTSxBujoEvzF564e8gt8.

"522nd Field Artillery Battalion." *Densho Encyclopedia*, https://encyclopedia.densho.org/522nd_Field_Artillery_Battalion/

"522 Liberates Dachau Prisoners." *Nisei Veterans Legacy*, https://www.nvlchawaii.org/522- . liberates-dachau-prisoners

Fischer, Wolfgang C., ed. *German Hyperinflation 1922/23: A Law and Economics Approach*, https://en.wikipedia.org/wiki/Papiermark.

"Frederick the Great." *Wikipedia*, https://en.wikipedia.org/wiki/Frederick_the_Great

Gallagher, Robert F. *World War II Story*, "Chapter 12 — Camp Stapley, England." (Robert F. Gallagher, 1999), https://gallagherstory.com/ww2/

"German Ten marks banknote." *Powerhouse Collections*, https://collection.maas.museum/object/390060

"German WW2 Luftwaffe Qualified Radio Operator Trade Patch." Warstuff, https://www.warstuff.com/product/german-ww2-luftwaffe-qualified-radio-operator-trade- patch/

Gershon, Livia. "When the Nazis Murdered Thousands by Sending Them on Forced Death Marches." *Smithsonian Magazine,* May 18, 2021, https://www.smithsonianmag.com/smart-news/remembering-nazi-death-marches-180977751/

Goodell, Nick. "1920s Hyperinflation in Germany and Bank Notes." *University of Illinois Urbana-Champaign College of Liberal Arts and Sciences Museum of World Cultures,* https://www.spurlock.illinois.edu/blog/p/1920s-hyperinflation-in/283

"Hohenlohe — A German Princely Dynasty." *Wikiwand,* https://www.wikiwand.com/en/Hohenlohe .

"Hitler Youth — Hitler's Boy Soldiers." *The History Place,* https://www.historyplace.com/worldwar2/hitleryouth/hj-boy-soldiers.htm

"Hitler Youth Uniforms: DJ Rune (Lighting Bolt) Patch." *Historical Boys' Uniforms,* https://histclo.com/youth/youth/org/nat/hitler/uni/insig/ip-bolt.htm

"Iron Cross." *Britannica,* August 31, 2022, https://www.britannica.com/topic/Iron-Cross

Johnson, Colonel Gerdon F. *History of the Twelfth Infantry Regiment in World War II,* 2nd Ed. (1991). Boston: National Fourth (Ivy) Division, 1947.

Klinek, Eric William. *The Army's Orphans: The United States Army Replacement System In the European Campaign,* 1944-1945 (Dissertation Submitted to the Temple University Graduate Board, May, 2014).

Lerwill, Lieutenant Colonel Leonard. *The Personnel Replacement System in the United States Army* (Dept of the Army, 1954).

"Liberty Ship." *Wikipedia*, https://en.wikipedia.org/wiki/Liberty_ship

MacDonald, Charles C., quoted in Neillands, Robin (2005). *The Battle for the Rhine 1945.* London: Orion Publishing Group, https://en.wikipedia.org/wiki/Battle_of_Hürtgen_Forest

Memo: Headquarters 12th Infantry APO 4, US Army, Subject: Action Against Enemy, Reports After/After Action Reports. To: The Adjutant General, Washington, 25 D.C. 4 January 1945.

Miles, Ralph J. *War Album, Volumes 1 and 2.* Personal collection of David W. Miles, Moody, Alabama

"Military History of Rothenburg ob der Tauber." *Fandom,* https://military-history.fandom.com/wiki/Military_history_of-rothenburg_ob-der-tauber

"National Socialist Flyers Corps." Military-history.fandom.com, https://military.wikia.org/wiki/National_Socialist_Flyers_Corps.

"Nazi Germany." *Wikipedia,* https://en.wikipedia.org/wiki/Nazi_Germany

Prefer, Nathan N. "Bloodletting in the Hurtgen Forest." *Warfare History Network,* warfarehistorynetwork.com

"1,000 Mark Reichsbankbnote Imperial Germany Circa 1910." *Collectors Weekly,* https://www.collectorsweekly.com/stories/225099-1000-mark- reichsbanknote-imperial-germ

"Original German NASDP Enamel Membership Badge Pin." *IMA-USA.com,* https://www.ima- usa.com/products/original-german-nsdap-party-enamel-membership-badge-pin-by-karl-hensler-rzm-6?variant=31668433158213

"Original German WWII Deutsche Wehrmacht Armband." *1944 Militaria,* https://www.1944militaria.com/Original_German_WWII_Deutsche_Wehrmacht_Armband_p/ordwamb.htm

"Photos Then." *IM Westen,* http://imwesten.com/photos-then/

Record Group 407; File 304-INF (12). Office of the Adjutant General. National Archives and Records Administration, College Park, MD.

Rieken, John. "Fort McPherson." *New Georgia Encyclopedia,* July 19, 2022 at https://www.georgiaencyclopedia.org/articles/government-politics/fort-mcpherson/

Rundell, Walter. "Currency Control by the United States Army in World War II: Foundation for Failure," *Pacific Historical Review* 30:4, https://azmemory.azlibrary.gov/digital/collection/msiartifact/id/14/.

"Schloss Schillingsfurst." *Second Wiki,* https://second.wiki/wiki/schloss_schillingsfc3bcrst

Schwartz, E. (2011). The Death Marches from the Dachau Camps to the Alps during the Final Days of

World War II in Europe. Dapim: Studies on the Holocaust, 25(1), 129–160. https://doi.org/10.1080/23256249.2011.10744408

Shores, Max. "Up From the Ashes: The Rebirth of Phenix City," https://maxshores.com/up-from-the-ashes-the-rebirth-of-phenix-city/

"Stamps of Germany: World Jamboree of Hitler Youth Issues (1935)." *The Philatelic Database*, http://www.philatelicdatabase.com/hitler/stamps-of-germany-world-jamboree- of-hitler-youth-issues-1935/

The Ardennes/The Battle of the Bulge, "Chapter X: The German Southern Shoulder is Jammed." *US Army Center of Military History*, https://www.history.army.mil/index.html

"The Battle of the Hurtgen Forest." *Liberation Route Europe,* https://www.liberationroute.com/themed-routes/5/the-battle-of-the-hurtgen-forest

"The Battle of the Huertgen Forest — Schlacht im Hurtgenwald." http://home.scarlet.be/~sh446368/aar-12th-inf-1.html

"The Drive on Prum 29 January — 14 February 1945." *IM Westen*, https://imwesten.com/drive-on-pruem/

"The 4th Infantry Division During World War II." *Holocaust Encyclopedia,* https://encyclopedia.ushmm.org/content/en/article/the-4th-infantry-division

"The Hurtgen Forest Project ." *Defense POW/MIA Accounting Agency,* https://www.dpaa.mil/Portals/85/Briefing%20Video%20Files/Huertgen_Project_Slides.p.df?ver=PEN_4b_yW2sEEUVTOYp-dAA%3D%3D

"The Riddle of the Hindenburg Medallions." https://www.rfrajola.com/PDR2017/Coffey1/Coffey1.pdf

"The Second World War: European Theater — Invasion of Germany Crossing of the Rhine." *United States Military Academy*, https://www.westpoint.edu/academics/academic-departments/history/world-war-two-europe

"Third Reich Commemorative Issues 1933." *Stamp-Collecting-World*, https://www.stamp- collecting-world.com/thirdreich_nothilfe.html

"Third Reich Commemorative Issues 1934." *Stamp-Collecting-World*, https://www.stamp- collecting-world.com/thirdreich_1934.html

"Third Reich Commemorative Issues 1935." *Stamp-Collecting World*, https://www.stamp-collecting-world.com/thirdreich_1935.html

"Third Reich Commemorative Issues 1943." *Stamp-Collecting-World*, https://www.stamp- collecting-world.com/thirdreich_1943.html

"Third Reich Definitive Issues." *Stamp-Collecting-World,* https://www.stamp-collecting-world.com/thirdreich_hitlerdef.html

"Third Reich — World Olympic Games 1936." *Stamp-Collecting-World*, https://www.stamp-collecting-world.com/thirdreich_olympics.html

"36th Division in World War II — Prize Catch." *Texas Military Forces Museum.com,* http://www.texas-militaryforcesmuseum.org/36division/archives/seigfri/prize.htm

"Training the GI." https://www.nationalww2museum.org/war/articles/training-american-gi

"Volkssturm." *Wikipedia*, https://en.wikipedia.org/wiki/Volkssturm.

"Walter Donicke." *Wikipedia,* https://de.wikipedia.org/wiki/Walter_Dönicke

"Weimar Republic Hyperinflation Issues 1923." *Stamp-Collecting-World*, https://www.stamp- collecting-world.com/weimarrepublic_hyperb.html

"Weimar Republic." Wikipedia https://en.wikipedia.org/wiki/Hyperinflation_in_the_Weimar_Republic

Weiner, Leslie. "SS Sea Bass — General Orders," posted on FB.com at National Fourth Infantry Division Association, July 23, 2020.

"What Is a US ASR Score?" *Chron.,* March 11, 2021, https://work.chron.com/army-asr-score-22134.html

"What is the value of a 1000 Mark Reichsbanknote from 1910?" *Leftover Currency*, February 18, 2017, https://www.leftovercurrency.com/1000-mark-reichsbanknote-1910-value/

"WWII — Coins, banknotes and collections in various packages related to the Second World War." *Educational Coin Company*, https://www.educationalcoin.com/media/amfile/files/(1)imageshistoryfolio-salliedmiltary8bnfolio.pdf

"World War II Militaria." *Gettysburg Museum of History,* https://www.gettysburgmuseumofhistory.com/gettysburg-battle/world-war-ii- militaria

Zargar, Major Glenn, Military Monograph: *The Defense of Little Switzerland.* Carlisle, PA: United States Military History Institute, May 1, 1948.

www.ingramcontent.com/pod-product-compliance
Lightning Source LLC
Chambersburg PA
CBHW081324120626

46546CB00011B/3208